Stable and Random Motions
in Dynamical Systems

PRINCETON LANDMARKS
IN MATHEMATICS AND PHYSICS

Stable and Random Motions in Dynamical Systems: With Special Emphasis on Celestial Mechanics, *by Jürgen Moser*

PCT, Spin and Statistics, and All That, *by Raymond F. Streater and Arthur S. Wightman*

Non-Standard Analysis, *by Abraham Robinson*

General Theory of Relativity, *by P.A.M. Dirac*

Angular Momentum in Quantum Mechanics, *by A. R. Edmonds*

Mathematical Foundations of Quantum Mechanics, *by John von Neumann*

Introduction to Mathematical Logic, *by Alonzo Church*

Convex Analysis, *by R. Tyrell Rockafellar*

Riemannian Geometry, *by Luther Pfahler Eisenhart*

The Classical Groups: Their Invariants and Representations, *by Hermann Weyl*

Topology from the Differentiable Viewpoint, *by John W. Milnor*

Algebraic Theory of Numbers, *by Hermann Weyl*

Continuous Geometry, *by John von Neumann*

Linear Programming and Extensions, *by George B. Dantzig*

Operator Techniques in Atomic Spectroscopy, *by Brian R. Judd*

The Topology of Fibre Bundles, *by Norman Steenrod*

Mathematical Methods of Statistics, *by Harald Cramér*

Theory of Lie Groups, *by Claude Chevalley*

Homological Algebra, *by Henri Cartan and Samuel Eilenberg*

# STABLE AND RANDOM MOTIONS IN DYNAMICAL SYSTEMS

*With Special Emphasis on Celestial Mechanics*

BY

JÜRGEN MOSER

WITH A NEW FOREWORD BY

PHILIP J. HOLMES

Hermann Weyl Lectures
The Institute for Advanced Study

PRINCETON UNIVERSITY PRESS
PRINCETON AND OXFORD

Published by Princeton University Press, 41 William Street,
Princeton, New Jersey 08540
In the United Kingdom: Princeton University Press, 3 Market Place,
Woodstock, Oxfordshire OX20 1SY



Originally published in association with the University of Tokyo Press in 1973,
as part of the Annals of Mathematics Studies, No. 77

Third paperback printing, and first Princeton Landmarks in Mathematics edition,
with a new foreword by Philip J. Holmes, 2001

**Library of Congress Cataloging-in-Publication Data**

Moser, Jürgen, 1928–1999.
Stable and random motions in dynamical systems : with special emphasis on celestial mechanics / by Jürgen Moser ; with a new foreword by Philip J. Holmes.
p. cm. — (Princeton landmarks in mathematics and physics)
Originally published: Princeton, N.J. : Princeton University Press, 1973.
Includes bibliographical references.
ISBN 0-691-08910-8 (pbk. : alk. paper)
1. Celestial mechanics. I. Title. II. Series.

QB351 .M74 2001
521—dc21

2001016309

British Library Cataloging-in-Publication Data is available

The original work was partially supported by the grant AFOSR-71-2055

Printed on acid-free paper. ∞

www.pup.princeton.edu

3 5 7 9 10 8 6 4

# HERMANN WEYL LECTURES

The Hermann Weyl lectures are organized and sponsored by the School of Mathematics of the Institute for Advanced Study. Their aim is to provide broad surveys of various topics in mathematics, accessible to nonspecialists, to be eventually published in the Annals of Mathematics Studies. It is intended to have one or two such sets of lectures each year.

The present monograph is the first one in this series. It is an outgrowth of five lectures given by Professor J. Moser at the Institute on February 7, 14, and 21, 1972.

ARMAND BOREL
JOHN W. MILNOR
1973

index theory, linearization, and invariant manifolds. He summarized most of this in the three volumes of *Les méthodes nouvelles de la mécanique céleste* [E], and the lesser-known (and never translated) *Leçons de mécanique céleste* [F].

Development of these ideas was rather slow, but G. D. Birkhoff (eg. [G]) in the United States and A. A. Andronov and L. S. Pontryagin in the Soviet Union (eg. [H]) made significant contributions, followed by those of D. V. Anosov and S. Smale (eg. [I]). It was Smale who, in 1959–60, essentially completed the analysis of orbits passing near homoclinic points that Poincaré had begun in [B], and to which Birkhoff had also contributed. Using symbolic dynamics, Smale provided a complete description of the prototypically chaotic set (albeit of "saddle-type," and hence not an attractor) that is found near transverse homoclinic points in differentiable mappings. Ironically, Smale's paper describing this work did not appear until 1965 [J], and then in a form obscure for many dynamical systems researchers. While a summary of the ideas was contained in [I], the treatment found in Chapter III of the present book was the first complete and accessible account, including explicitly computable conditions that enable one to verify the presence of a hyperbolic structure, and a fully-worked example. Coupled with V. K. Melnikov's method [K] for locating homoclinic points in periodically perturbed ordinary differential equations, this now provides a powerful set of tools for proving and analyzing chaotic motions in specific nonlinear systems. Here the method is applied to a version of the three-body problem which is different from that considered by Poincaré, and results of Sitnikov and Alekseev are recovered (with details supplied in Chapter VI). This is the "random" thread in the present volume.

The "stable" thread has its origins further back in celestial mechanics, in the attempts of Laplace, Lagrange, Poisson, and others to prove stability of the solar system by perturbative means. This eventually led to normal form methods, also developed by Poincaré, in which successive (nonlinear) coordinate changes are used to simplify, term by term, an analytic Hamiltonian function expressed as a power series. If one simply neglects higher-order terms, then the dynamical behavior near elliptic fixed points for typical (nondegenerate) Hamiltonian systems is quasi-periodic: essentially a set of nonlinear normal modes constituting a flow on an invariant torus. Formally, the system appears to be completely integrable and expressible in action-angle coordinates, with all the actions remaining constant under the flow. However, Poincaré's work had shown that formal series representations of solutions failed to converge in certain cases.

Thus the difficulty lay in proving that, under suitable conditions, convergence *does* occur for a class of quasi-periodic solutions with sufficiently irrational frequency ratios, in which case the higher-order terms really can be neglected. This was achieved by V. I. Arnold, and by Moser in two independent sets of work that appeared in 1962–63 [L,M], following A. N. Kolmogorov's discov-

TABLE OF CONTENTS

## FOREWORD TO THE PRINCETON LANDMARKS IN MATHEMATICS EDITION

Jürgen Moser had a lifelong interest in celestial mechanics and dynamical systems theory. In Göttingen he studied for his doctorate under the direction of Franz Rellich, and when Carl Siegel returned in 1950, it was Rellich who arranged for Moser to write up the notes for Siegel's lectures on celestial mechanics. This led to the classic *Lectures on Celestial Mechanics*, which appeared in German in 1956, in English translation in 1971, and in reprint in 1995 [A]. The present volume is a second, lesser-known classic, and Moser's first major pedagogical work as sole author. As the original note by the series editors states, it derives from a series of five Hermann Weyl lectures given by Moser at the Institute for Advanced Study in Princeton, New Jersey, in February 1972 (it was in fact the first in that series). If it has not had the impact of [A], this is perhaps due to its comparatively small initial print run. Most appropriately, it is now being reprinted in Princeton's Landmarks in Mathematics series, for it represents an important step in the development of the qualitative theory of dynamical systems.

The two strands making up Moser's book are stability and randomness, or chaos. The main results described are the Kolmogorov-Arnold-Moser (KAM) theorem, which proves that many regular, quasi-periodic motions survive in near-integrable Hamiltonian systems, and the "Smale-Birkhoff" homoclinic theorem, which establishes equivalence between a set of orbits and a Bernoulli process. The following brief historical sketch is intended to set the scene and help the reader place and better appreciate Moser's contribution.

The modern geometrical theory of dynamical systems originated between 1880 and 1910 in Poincaré's work on ordinary differential equations and celestial mechanics, and particularly in his prize-winning essay on Hamiltonian mechanics and the three-body problem [B], in which he discovered homoclinic and heteroclinic orbits (solutions doublement asymptotiques) and thereby began to appreciate the phenomenon of deterministic chaos. The fascinating story of this work, the mistake contained in Poincaré's original submission to the prize jury, and his subsequent correction of the paper *after* the prize (celebrating the $60^{\text{th}}$ birthday of King Oscar I of Sweden and Norway) had been awarded, is told in [C,D]. In [B] and a series of earlier papers in *Comptes Rendus* and *J. de Math. pures et appl.*, Poincaré developed basic tools such as the return (Poincaré) map, ery and announcement of the main ideas, contained in his lecture at the Amsterdam International Congress of Mathematicians [N]. (There is an interesting story here, too, regarding Moser's contribution, his interest having been first sparked when he was asked to summarize Kolmogorov's announcement for Mathematical Reviews. However, he was unable to fill in all the details, and this evidently prompted him to begin working on the problem himself: see [D].) The material on normal forms, a version of the Kolmogorov-Arnold-Moser theorem on persistence of invariant tori (Theorem 2.9), and Moser's Twist theorem for area-preserving maps form Chapter II of this book, and the technical details concerning small divisors are given in Chapter V.

The balance of the book is Chapter I—an introduction to celestial mechanics and the stability problem, which contains some motivating examples—and the brief concluding remarks in Chapter IV. This chapter ends by mentioning the Fermi-Pasta-Ulam system and certain completely integrable partial differential equations, including that of Korteweg-deVries. Moser remarks that "the question [of] whether some results of the finitely many degrees of freedom [case] can be carried over [to PDEs] . . . seem[s] rather speculative . . . ." Subsequently, Moser himself made significant contributions to this question, and it remains a vital research area today. That is, however, another story. (An article on Moser's life and mathematical contributions, written by his former colleagues and students, appears in [O].)

Shortly before his untimely death in December 1999, Jürgen Moser had endorsed the reissue of this book. It makes a fitting memorial to a great mathematician and teacher. After almost thirty years, Moser's lectures are still one of the best entrées to the fascinating worlds of order and chaos in dynamics.

PHILIP J. HOLMES
Princeton University
Fall 2000

## REFERENCES

[A] SIEGEL, C. L. and MOSER, J. K., *Lectures on Celestial Mechanics*, Springer Verlag, Berlin, 1971 (reprinted in the Classics in Mathematics series, 1995).

[B] POINCARÉ, H. J., Sur le problème des trois corps et les équations de la dynamique, *Acta Mathematica* 13 (1890) 1–270.

[C] BARROW-GREEN, J., *Poincaré and the Three Body Problem*, American Mathematical Society and London Mathematical Society, AMS Press, Providence, RI, 1997.

[D] DIACU, F. and HOLMES, P., *Celestial Encounters: The Origins of Chaos and Stability*, Princeton University Press, Princeton, NJ, 1996.

[E] POINCARÉ, H. J., *Les méthodes nouvelles de la mécanique céleste,* vols. I–III, Gauthiers-Villars, Paris, 1892, 1893, 1899 (reprinted by Librarie Albert Blanchard, Paris, 1987; and in English translation with a preface by D. Goroff, American Institute of Physics Press, New York, 1993).

[F] POINCARÉ, H. J., *Leçons de mécanique céleste,* Gauthiers-Villars, Paris, 1905.

[G] BIRKHOFF, G. D., *Dynamical Systems,* American Mathematical Society, AMS Press, Providence, RI, 1927 (reprinted with an introduction and addendum by J. K. Moser and a preface by M. Morse, 1966).

[H] ANDRONOV, A. A. and PONTRYAGIN, L. S., Coarse systems, *Dokl. Akad. Nauk. SSSR* 14 (1937) 247.

[I] SMALE, S., Differentiable dynamical systems, *Bull. Amer. Math. Soc.* 73 (1967) 747–817.

[J] SMALE, S., Diffeomorphisms with many periodic points, In *Differentiable and Combinatorial Topology,* S. S. Cairns (ed.), pp. 63–80, Princeton University Press, Princeton, NJ, 1965.

[K] MELNIKOV, V. K., On the stability of the center for time periodic perturbations, *Trans. Moscow Math. Soc.* 12 (1963) 1–57.

[L] ARNOLD, V. I., Proof of A. N. Kolmogorov's theorem on the preservation of quasi-periodic motions under small perturbations of the Hamiltonian, *Russ. Math. Sur.* 18 (5) (1963) 9–36.

[M] MOSER, J. K., On invariant curves of area-preserving mappings in the plane, *Nachr. Akad. Wiss. Göttingen II, Math. Phys. Kl.* (1962) 1–20.

[N] KOLMOGOROV, A. N., Théorie générale des systèmes dynamiques et mécanique classique, In *Proceedings of the International Congress of Mathematicians,* Amsterdam, 1954, vol. 1, pp. 315–333, North-Holland Publishing Co., Amsterdam, 1957.

[O] MATHER, J. N., MCKEAN, H. P., NIRENBERG, L., and RABINOWITZ, P. H., Jürgen K. Moser (1928–1999), *Notices of the Amer. Math. Soc.* 47 (11) (2000) 1392–1405.

Stable and Random Motions
in Dynamical Systems

# CHAPTER I

# INTRODUCTION

## 1. *The stability problem*

### a) *The* N-*body problem*

Celestial Mechanics is a subject with a tremendously long and varied history and it is clearly impossible to give a survey of this field in a short space. Our goal will be just to report on some progress in the last few decades, and even there we will be very selective. In a very simplified manner one may describe the goal of Celestial Mechanics as the study of the solutions of one system of differential equations, namely the N-body problem dealing with the motion of N masspoints in the three-dimensional space attracting each other according to Newton's law. If $x_k$, $k = 1,2,\dots,N$ denote N three-dimensional vectors describing the position of N masspoints of positive mass $m_k$ this system of differential equations has the familiar form

$$m_k \frac{d^2 x_k}{dt^2} = \frac{\partial U}{\partial x_k}, \qquad k = 1,2,\dots,N \tag{1.1}$$

where

$$U = \sum_{1 \le k < \ell \le N} \frac{m_k m_\ell}{|x_k - x_\ell|} \tag{1.2}$$

and $|x_k - x_\ell|$ denotes the Euclidean distance. Although the solutions to this problem can be given explicitly for $N = 2$ very little is known about it if $N \ge 3$, in spite of the efforts of many astronomers and mathematicians who attacked this problem on account of its importance for astronomy.

In the early stages of this field its task was quite pragmatically to calculate the orbits of the planets and predict the ephemerides over a fairly long time. This required mathematical tools and was largely what now would be called numerical analysis. This problem, of course, still has its important place, in which nowadays computing machines play an important role. However, the mathematical difficulties connected with this field inspired more and more the study of basic theoretical problems leading to the development of new mathematical tools. The main influence in this direction came from Poincaré who started a number of new lines of thought, like the qualitative theory of differential equations, the quest for the topology of the energy manifold, he formulated and proved fixed point theorems to establish the existence of periodic solutions. In this connection one may recall that the fixed point theorem by P. Bohl and Poincaré also had its origin in this field, to which Bohl devoted his entire mathematical life. As a consequence, in the beginning of this century the more basic questions of theoretical nature attracted the attention of mathematicians, like Bohl, Liapunov, Sundman, Cherry, and in more recent times Siegel, Kolmogorov, Arnold and others. Only with the last part of this development will these lectures be concerned, and mainly with the recent advances on the stability problem (Chapter II) and, in the opposite direction, with some orbits whose behavior is quite erratic or random (Chapter III).

b) *Stability problem*

There are many different formulations of the stability problem which explains that there are so many proofs of the stability of the solar system, like those of Laplace, Lagrange, Poisson, etc. [12], [13]. We mention, rather informally, some basic questions: Looking at the differential equations (1.1) and (1.2) one notices that they are defined only for $|x_k - x_\ell| \neq 0$ $(k < \ell)$ and if $r_{k\ell} = |x_k - x_\ell|$ approaches zero we speak of a collision. Thus the second order system of differential equations (1.1) is defined in the complement of the $\frac{1}{2}N(N-1)$ hyperplanes $x_k = x_\ell (k < \ell)$ in the 3N-dimensional Euclidean space. The main problem is the study of the behavior of the solutions for an infinite time interval. Are there solutions which do not experience collisions and do not escape? Are there solutions for which

$$\max_{k<\ell} (r_{k\ell}, r_{k\ell}^{-1})$$

is bounded for all $t$? These questions can be answered by the construction of periodic solutions. But, are these solutions exceptional, or do they form an open set or at least a set of positive measure, in the phase space? These questions are motivated by observations in astronomy, although one has to keep in mind that the observations cover a finite and rather limited time interval, that in the above formulation all other possible forces, including relativistic effects, have been neglected and the problem really is to be considered a mathematical idealization.

As a rule one considers the N-body problem only in the case where $N-1$ mass ratios, say $m_k/m_N (k = 1,2,\ldots,N-1)$ are small, the planetary problem. Normalizing $m_N = 1$ (mass of sun) we treat $m_1,\ldots,m_{N-1}$ as small parameters. If one divides the $k^{th}$ equation (1.1) by $m_k$ and goes to the limit $m_k \to 0$ for $k = 1,2,\ldots,N-1$ the differential equations decouple, as is also intuitively clear, and describe $N-1$ two-body problems, for the $N-1$ planets. If we choose the center of mass at rest and for the $N-1$ planets elliptical orbits with the frequencies $\omega_k (k = 1,\ldots,N-1)$, i.e., with periods $2\pi/\omega_k$, we obtain a solution which will in general not be periodic, but quasi-periodic. Here we call a vector function quasi-periodic if its components can be represented by a series of the form

$$\sum_j c_j e^{i(j,\omega)t} \tag{1.3}$$

where $\omega = (\omega_1, \omega_2, \ldots, \omega_s)$ has real components and the summation is taken over all vectors $j = (j_1, j_2, \ldots, j_s)$ with integer components. The sum

$$\sum |c_j|$$

is assumed to be convergent. In our case we have $s = N-1$ and the question arises whether one can find such quasi-periodic solutions for small but positive values of $m_k (k = 1,2,\ldots,N-1)$ for the planetary problems.

This question for such quasi-periodic solutions is a very old one. In a letter to S. Kovalevski in 1878 (see [14], p. 30) Weierstrass mentioned that he was able to formally construct such series solutions but that he was unable to establish their convergence. The difficulty of the convergence of such series is connected with the notorious occurrence of the small divisors. Roughly the difficulty is the following: The analytic formalism breaks down if the frequencies $\omega_k$ are rationally dependent since the expressions

$$\sum_k j_k \omega_k \tag{1.4}$$

enter into the denominator of the coefficients $c_j$. Therefore one is forced to assume that the $\omega_k$ are rationally independent, but then the expressions (1.4) come arbitrarily close to zero and this is the reason for the term "small divisors." Since these expressions (1.4) enter into the coefficients in a very complicated manner the convergence of the series becomes very doubtful. This has caused several people to doubt the stability of the solar system. The situation is further compounded since one knows from the study of resonance that one may construct unstable systems for rationally related frequencies, a phenomenon which occurs for Saturn and Jupiter with a frequency ratio 2/5. Therefore some people suspected that rational dependence of the frequencies is related to the question of stability of the planetary system which is absurd from the physical point of view since any set of frequencies can be approximated by rationally dependent ones!

This confused state of affairs is illustrated by the following comment by Weierstrass translated from a lecture, held in Berlin 1880-81 [15]:

"... Thus we see that the stability cannot depend on rationality or irrationality of certain quantities. Precisely this error has been committed by Biot. Ever since this mistake has been made it is spread in big lectures over the organization of the world. He said, a slight change in the distance between Saturn and Jupiter would be sufficient that the strangest of all planets in our system would escape our system forever. However, one forgot to mention that also Jupiter could escape, and this would indeed simplify the work of astronomers considerably, since this is precisely the planet causing the largest perturbation.''

On the other hand, there are several configurations in the solar system for which the frequencies are rationally related. We mentioned already such a phenomenon for Saturn and Jupiter; the frequency ratio of Uranus and Neptune is close to 2 and for the frequencies of Jupiter, Mars and Earth the expression

$$3\omega_J - 8\omega_M + 4\omega_E$$

is approximately zero. Between three of the Galilean moons of Jupiter one has a relation

$$\omega_I - 3\omega_{II} + 2\omega_{III} \sim 0$$

to high degree of accuracy. There are recurring controversies whether or not these phenomena can be viewed as statistical accidents (see, for example, [16]-[21]), since for 9 real numbers it is always possible to fulfill a certain number of rational relations to some degree of accuracy. For the asteroids, however, a rational relation with small integers of their frequencies with that of Jupiter seems to be exceptional.

All these different observations and speculations make it desirable to have a definitive existence theorem of quasi-periodic solutions for the N-body problem.

This problem was solved only in the last decade. The successful solution of the convergence problem connected with the small divisors is due to Siegel, Kolmogorov, Arnold and the author. Their work permits applications to other classes of differential equations like holomorphic and Hamiltonian systems (Chapter II). But, as far as the planetary problem is concerned these methods yield the result that for sufficiently small $m_1, m_2, \ldots, m_{N-1}$ there do indeed exist quasi-periodic solutions with $s = 3(N-1)$ rationally independent frequencies. Moreover, these orbits form a set of positive measure in the phase space. This theorem by

Arnold [22] answers an old problem and establishes not only the existence of quasi-periodic solutions for the N-body problem, but shows that these solutions are not exceptional in the sense of Lebesgue measure. It provides a set of positive measure of rigorous solutions which avoid collisions and infinity for all time and one may consider this result as a stability statement, although the set of orbits so obtained, which are nearly circular and have their inclination near zero, do not form an open set but a complicated Cantor set.

## 2. *Historical comments*

At this point it may be appropriate to insert some historical remarks. One finds some interesting information about the problem of the existence of quasi-periodic solutions in the letters of Weierstrass, published in the Act. Math. 35, 1911, pp. 29-65. As was mentioned already, Weierstrass had a great interest in the problem of constructing such solutions for the N-body problem, and he was in the possession of formal series expansions for such solutions, as he expressed in a letter to S. Kovalevski. He attempted to prove the convergence of these series expansions which contain the above mentioned small divisors. His hope for overcoming this difficulty was based largely on a remark made by Dirichlet to Kronecker in 1858 that he had found a method to approximate the solutions of the N-body problem successively. Since Dirichlet died soon afterwards there were no written notes along these lines available. Weierstrass apparently interpreted this remark in the sense that the above series expansion are, in fact, convergent. Later he suggested this problem to Mittag-Leffler as a prize question sponsored by the Swedish king (see [14]). He read and evaluated the famous paper by H. Poincaré [23] to whom the prize was awarded. Actually this paper did not contain the solution to the problem posed but contained a wealth of new ideas. As far as this problem is concerned, however, Poincaré went out to prove that the three-body problem does not possess any integrals aside from the ten known ones and functions of these. He indicated that this implied nonexistence of quasi-periodic solutions which was contrary to Weierstrass expectations. This argument, however, was not conclusive and Weierstrass criticized this seemingly fussy point, by writing:

"Nun habe ich noch ein Desiderium, worüber ich selbst an P. schreiben werde.

P. behauptet, dass aus der Nichtexistenz mehrerer eindeutigen (analytischen) Integrale bei einem dynamischen Probleme nothwendig die Unmöglichkeit folge, das Problem durch Reihen von der Form

$$\sum C_{\nu\nu'\ldots} \frac{\cos}{\sin} (\nu a t + \nu' a' t + \ldots)$$

zu lösen. Diese Behauptung, die von fundamentaler Bedeutung ist, wird ohne Beweis ausgesprochen. ..."

If one follows the later presentations of Poincaré on this point it is indeed discussed with great care. Most important in this connection is Section 149 of his Méthodes Nouvelles II [24] where he discusses the possibility of convergence or the divergence of his series expansions for the case of appropriately chosen fixed frequencies (independent of the parameter $\mu$). His assertion about this case is noncommittal and he wrote: "This seems to allow us to conclude that the above series (2) do not converge. Nevertheless the above argument does not suffice to establish this point with all rigor."

With the work of Kolmogorov and Arnold we know that, in fact, the opposite is the case and that these series expansions do converge and represent bona fide solutions of the problem, at least if a certain Hessian determinant does not vanish.* Thus we can say that Weierstrass' question is finally answered in the positive sense. Moreover, this is not in contradiction to Poincaré's theorem on the nonexistence of integrals. Whether this is related in any way to Dirichlet's approach can, of course, only be speculation.

---

* This refers to the condition (3.10) of Chapter II.

3. *Other problems*

The above mentioned results concerning quasi-periodic solution apply, of course, not only to the N-body problem but other classes of differential equations. As a rule it is of importance that the differential equations admitted describe motion without dissipation, as is the case with the planetary motion. More generally, the equations of mechanics are written as Hamiltonian systems

$$\dot{x}_k = \frac{\partial H}{\partial y_k}, \quad \dot{y}_k = -\frac{\partial H}{\partial x_k} \qquad (k = 1,2,\dots,n)$$

where H is a function of $x_1,\dots,x_n$, $y_1,\dots,y_n$, and the equations (1.1) obviously can be put into this form. Another class of admissible systems are of the form

$$\frac{d^2x_k}{dt^2} = F_k(x) , \qquad k = 1,2,\dots,n$$

where $F_k$ depends on $x = (x_1,x_2,\dots,x_n)$. If $x(t)$ is a solution so is $x(-t)$ and this is a special case of so-called reversible system.

In the following we will not discuss the technically involved N-body problem but rather treat the stability problem for typical model problems, like Hamiltonian or reversible systems. We illustrate these problems with some examples of apparently simpler nature.

i) In mechanics one frequently encounters the Duffing equation

$$\ddot{x} + ax + bx^3 = p(t)$$

where $p(t) = p(t+2\pi)$ is a periodic forcing function. If $p = 0$ and a, b both positive constants it is well known that all solutions are periodic, with a period depending on the amplitudes. However, even if $p(t)$ is such that $\max_t |p(t)| > 0$ is small it is a very delicate problem to decide whether all solutions are bounded. This is indeed the case, as a result of the theory to be discussed.* In this case the free vibration and the forcing term give rise to a small divisor problem.

---

* Recently G. Morris established the boundedness of the solutions of the above equation for $a = 0$ without smallness restriction on $p(t)$. His proof also rests on Theorem 2.11.

A similar problem is the periodically excited nonlinear pendulum which belongs to the same category.

ii) We consider a geometrical problem in the plane:* Let $\Gamma$ be a closed oriented smooth convex curve in the plane with positive curvature. We define a mapping $\phi$ in the exterior $E$ of $\Gamma$ by drawing from any point $p \in E$ a tangent to $\Gamma$ and extend it to a point $q$ such that the distances from the point of contact to $p$ and $q$ are equal. To make the construction unique we select one of the two possible tangents in accordance with the orientation of $\Gamma$. Setting $\phi(p) = q$ we have defined a mapping $\phi$ of $E$ onto itself. If $\Gamma$ is twice continuously differentiable $\phi$ is a diffeomorphism of $E$ onto itself, and, as one proves easily, preserves the area element in the plane.

We study the iterates $\phi^k$ of $\phi$ where $\phi^1 = \phi$, $\phi^k = \phi^{k-1} \circ \phi$, $k = 2,3,\ldots$ . Similarly we denote by $\phi^{-1}$ the inverse mapping of $\phi$ and $\phi^{-k}$ the $k^{th}$ iterates of it, as well as $\phi^0$ the identity mapping. For any $p \in E$ we shall call the set $\{\phi^k(p),\ k = 0, \pm 1, \pm 2, \ldots\}$ the orbit through $p$. Such an orbit is obtained by repeated construction of tangents (see Fig. 1). A periodic orbit is determined by a point $p$ for which $\phi^n(p) = p$ for some positive number $n$, the period. It gives rise to a closed polygon.

The problem is to decide whether every orbit is bounded. If the boundary curve $\Gamma$ is a circle or an ellipse the images of a point will lie on a similar circle or ellipse and the answer to our question is trivially affirmative. But for an arbitrary convex curve this problem is not at all trivial since it also involves small divisor difficulties. As a consequence of Theorem 2.11 below one can show that every orbit is indeed bounded if $\Gamma$ is six times continuously differentiable. More is true, namely that there are many orbits whose points are dense on a $C^1$-curve, but generally not all orbits need lie on such curves. This simple geometrical problem illustrates clearly many of the phenomena and questions to be discussed. For example, the existence of infinitely many periodic orbits is a consequence of Birkhoff's fixed point theorem 2.10.**

---

* This problem was suggested by B. H. Neumann about 1960. See also: P. C. Hammer, Unsolved Problems, Proc. Symp. Pure Math. vol. VII, Convexity, Am. Math. Soc. 1963.

** This is related to Poncelet's Schliessungsproblem in geometry.

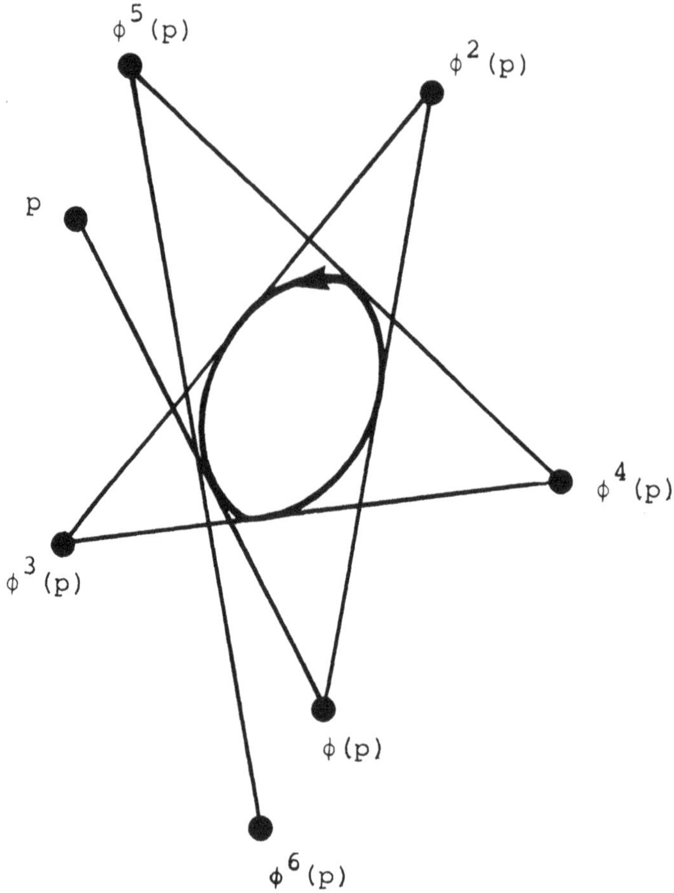


Fig. 1.

iii) If $\beta_k$, $k = 1,2,\ldots,n$, $n \geq 2$ are positive numbers we consider the system of differential equations

$$\frac{d^2 y_k}{dt^2} + \beta_k^2 y_k = Q_k(y) , \qquad k = 1,2,\ldots,n$$

where $Q_k(y)$ are smooth functions vanishing with their first derivatives if $y = (y_1,\ldots,y_n)$ is zero. Thus $y = 0$ is an equilibrium solution and, if all $Q_k$ were zero the equations describe n harmonic oscillators. The $Q_k$ can be viewed as weak coupling forces near $y = 0$ and the question is to decide about the stability of the equilibrium. Although this system is not Hamiltonian, in general, the methods of Chapter II are applicable to this case. Similarly, one may allow the $Q_k$ to depend periodically on the independent variable t; in the simplest case this leads to the equation

$$y'' + \beta^2 y = Q(t, y) ,$$

for which the stability question is not trivial but accessible to the methods of Chapter II, §4.

iv) The following mechanical problem was suggested by Ulam [25]. A particle is moving perpendicularly between two planes, one stationary and the other oscillating periodically along its normal. The particle is reflected elastically at the two walls. One may compare the situation with a player dribbling a basketball on the floor, but his hand moves periodically independent of the ball, except that the gravitation does not enter into our problem.

The question is to describe the motion of the particle over long times. Is it possible, say, that one can feed more and more energy into the particle by the periodic excitation so that its velocity can become arbitrarily large after a sufficiently long time? If the frequency of oscillation $\omega_w$ of the wall is large compared to that of the particle $\omega_p$ the phases of hitting the wall appear to be quite random. This led to the expectation that the velocity of the particle after a long time become quite arbitrary of the initial situation. Actually it turns out that the velocity remains bounded, and this is a consequence of Theorem 2.11, so that $\omega_p/\omega_w$ remain bounded also. If, however, the ratio is sufficiently small the motion is very erratic as was shown in a numerical study [26]. For large values of this ratio one finds many quasi-periodic motions.

There are a number of other situations to which the present theory can be applied. The motion of a charged particle in an electro-magnetic field can be described by a Hamiltonian system, and the special case of a charged particle in a magnetic dipole field like that of the earth is referred to as Störmer problems. Störmer and others studied the orbits for this problem by numerical calculations. Recently M. Braun [27], [28] established the existence of quasi-periodic motions for the Störmer problem, in which case the particle oscillates around the field lines and between the region near the north and south pole.

Another interesting example in mechanics is the motion of a gyroscope, or a spinning top and the stability problem connected with it [59].

4. *Unstable and statistical behavior*

a) *Ergodic theory*

So far we have described stable behavior of the solutions of a system of differential equations. However, in statistical mechanics one studies the opposite situation where the effect of the initial values gets wiped out after a long time. In particular, if one views a gas in a container as large number of particles interacting according to some force law one expects that after a sufficiently long time the positions of the particles are quite unrelated to the initial conditions. Various mathematical concepts describing such a random behavior are provided in ergodic theory, where one studies a group of measure preserving transformations $\phi^t(-\infty - t < +\infty)$ satisfying $\phi^{t+s} = \phi^t \circ \phi^s$ defined in a measure space. This group of transformations may be provided by a flow defined by a system of differential equations. In case of Hamiltonian systems the measure preserving property is a consequence of Liouville's theorem.

Similarly one considers a discrete group of measure preserving transformations $\phi^n (n = 0, \pm 1, \pm 2, \ldots)$ in a measure space. In this case this group is generated from $\phi = \phi^1$ by iteration and inversion.

According to the ergodic theorem (see, e.g., [29]) ergodicity implies that the solutions spend on the average as much time in a set $A$ as measure of $A$ indicates. That is, if the total space has measure $1$ and if $T_A$ is the time of a solution $\phi^t(p)$ spent in $A$ for $0 \leq t \leq T$ then

$$\frac{T_A}{T} \to m(A) \quad \text{as} \quad T \to \infty$$

for almost all p. In particular, if the measure space has a topology such that open sets are of positive measure then it follows that for an ergodic flow almost all orbits are dense in the entire space. Such a flow is called transitive. A simple example is the so-called Kronecker flow

$$\frac{d\theta_k}{dt} = \omega_k \, , \qquad k = 1,2,\ldots,s \tag{4.1}$$

on a torus, where $\theta_k$ are identified mod $2\pi$ and $\omega_1,\omega_2,\ldots,\omega_s$ are rationally independent real numbers. It is a consequence of a theorem of Kronecker that any solution of this system is dense on the torus if and only if the $\omega_1,\omega_2,\ldots,\omega_s$ are incommensurable. Clearly, the density of an orbit in the phase space precludes stability which requires that some orbits are restricted to an open subset of the phase space, and one may consider ergodicity or transitivity counterpoints to stability.

The ideas of statistical mechanics which are based on ergodicity of the considered flow suggested that the N-body problem may not be stable at all but even ergodic on some surface determined by the known integrals – at least if the number of particles is very large. That this is not so for the planetary motion is a by-product of Arnold's theorem mentioned already in Section 1 since the set of positive measure of quasi-periodic solutions is not compatible with ergodicity. Nevertheless there are classes of solutions whose behavior is quite random as we will see.

b) *Geodesic flow*

The question arises whether ergodicity, transitivity are typical phenomena for Hamiltonian systems of differential equations. For example, can ergodicity, or transitivity at least, persist under perturbation of the flow? The example of the Kronecker flow shows that this need not be the case, since the transitivity of the flow is connected with incommensurability of the $\omega_k$. On the other hand, there are other systems where transitivity of the flow persists under perturbation of the flow. Such flows were studied by Anosov [30] and are characterized by the property that differentially close orbits diverge from each other at an exponential rate. These systems have the property that all systems in a $C^1$-neighborhood of the given one have topologically the same orbit structure. That is, there exists a homeomorphism of the phase space taking the orbits of the first flow into those of the perturbed one. A flow with this property is called structurally stable. In particular, if a structurally stable system is transitive so are all nearby systems.

The concept of Anosov systems has its origin in the example of a geodesic flow on a two-dimensional compact manifold of negative curvature. These geodesics can also be viewed as the orbits of a masspoint moving on this manifold free from additional forces. The velocity of this masspoint is, of course, a constant and one may normalize it to be equal to 1. Thus we can take the three-dimensional unit tangent bundle of the manifold as the phase space of the flow. As was shown by Hedlund and Hopf (see [88]) this flow is indeed ergodic which implies that almost every orbit is dense in this 3-dimensional phase space.

These examples suggest that the ergodic behavior is the typical case for Hamiltonian flows and stability may be exceptional. This view was expressed by several authors and we just refer to a paper by E. Fermi [31], who at the time was interested in statistical mechanics and attempted a proof that a Hamiltonian system of $n$ degrees of freedom on some compact energy manifold is, in general, transitive on this $(2n-1)$-dimensional manifold. Also Birkhoff must have had this view as one can see from several remarks in his writing, for example [32].

That this view is erroneous is again a consequence of the existence theorems of quasi-periodic solutions to be described in Chapter II. One knows now that the presence of a single equilibrium solution which in linear terms is stable – and for which the nonlinear terms are not degenerate – implies the existence of non-trivial invariant sets (see Theorem 2.8), thus destroying ergodicity. The same is true if there is a periodic orbit which is stable according to the linearized theory and nondegenerate. Actually, that also transitivity is violated can be established only for two degrees of freedom at this time, and whether this is so for more degrees of freedom is an unsolved problem.

To compare the situation with the geodesic flow on a two-dimensional compact manifold we see from the familiar Jacobi equations that for negative curvature all periodic solutions are unstable and so there is no contradiction to our previous remark. If, however, the curvature is positive stable periodic solutions may exist. For example, any manifold close to a three-axial ellipsoid is of this nature and the geodesic flow in this case is neither ergodic nor transitive. However, many problems in applications do not have such strong instability behavior as in the case of negative curvature, that all periodic orbits are unstable and it is important to understand the essential feature of the flow in the case that also stable periodic orbits are present. This is the purpose of Chapter III where we will show that quasi-periodic behavior on a set of positive measure (stable behavior) and transitivity on other sets which are continua (unstable behavior) coexist side by side (see Theorem 3.9). In particular, we will discuss a beautiful example, studied by Sitnikov [33] and later by Alekseev [34,35], of a motion in the restricted three-body problem where a continuum of such solutions, called quasi-random by Alekseev, can be exhibited (see Chapter III, §5).

c) *Subsystems, sequence shift*

Since solutions of different behavior exist side by side it is important for the analytic and conceptual description to separate them and to study the flow on invariant subsets and invariant manifolds. For this purpose we make use of the concept of a subflow $\psi^t$ of a flow $\phi^t$. If $\phi^t$ is a smooth flow on a manifold M, say, and $\psi^t$ another flow on a manifold N we call $\psi^t$ a subsystem of $\phi^t$ if there exists a smooth embedding $w: N \to M$ of N in M taking the flow $\psi^t$ into the restriction of $\phi^t$ to $w(N)$. This concept is illustrated with the quasi-periodic solutions introduced earlier. They can be viewed as an embedding of the Kronecker flow into the given system.

Similarly one can consider merely continuous embeddings w of a continuum N into M for continuous subflows, and finally extend the concept to the discrete case of a group of mappings $\phi^n (n = 0, \pm 1, \pm 2, \ldots)$. To describe the random behavior of a mapping we will embed as subsystem of it the so-called sequence shift $\sigma$: If A is a finite or infinite set of elements $a \in A$ we introduce a topological space S whose elements s are doubly infinite sequences

$$s = (\ldots, s_{-1}, s_0; s_1, s_2, \ldots)$$

of elements $s_k$ in A. The topology in S is defined in such a way that two elements s, s′ are close if $s_k = s'_k$ for $|k| < K$ for a large number K. On this topological space we consider the homeomorphism $\sigma$ defined by $\sigma(s)_k = s_{k-1}$, which shifts the sequence by one notch to the right.

This sequence shift is the model for "quasi-random" behavior in the following and we will topologically embed it into a mapping connected with the restricted three-body problem. These embeddings of the sequence shift $\sigma$ which we will discuss also persist under perturbations of the system, but remarkably here all $C^1$-perturbations are permitted, not only conservative ones. This is in contrast to the situation for quasi-periodic solutions which will persist generally only if the nearby flows were restricted to Hamiltonian systems. Thus the quasi-random solutions appear more "stable under perturbation" than the quasi-periodic solutions.

## 5. *Plan*

In the following Chapter II we discuss existence theorems for quasi-periodic solutions and some applications in celestial mechanics. In Chapter III we treat the statistical behavior of the solutions where Sitnikov's and Alekseev's work on the restricted three-body problem plays the central role.

After some concluding remarks in Chapter IV we supply in the final two chapters some proofs. Chapter V contains the convergence proof for a typical small divisor problem and a discussion of the method, plus several generalizations of theorems in Chapter II. Similarly, Chapter VI contains proofs to theorems of Chapter III, in particular, a complete proof of Alekseev's result that one can continuously embed the sequence shift, into a mapping naturally derived from the restricted three-body problem, and the nonexistence of an integral for this problem. Thus Chapters II and V, as well as Chapters III and VI belong together.

There are a number of recent results which have been omitted. For example, Smale's work [36] on the topology of the manifold obtained by fixing the energy and angular momentum integrals for the N-body problem gives rise to the existence proofs of similarity solutions. There is a similar study due to Easton [37].

We also omitted various methods of finding periodic solutions, for which Morse theory and other topological approaches can be used (see [40], [41]). We just refer to a recent interesting paper by A. Weinstein [38, 39]. A study of generic properties of Hamiltonian systems is due to C. Robinson [42].

The problems of Celestial Mechanics bridge over to a number of other fields. In particular, it has contact with ergodic theory where many remarkable advances have been made. We mentioned the work of Anosov [30] and have to add Sinai's work [43] which is more closely related to statistical mechanics. As far as the metric equivalence of Bernoulli shifts is concerned we mention the remarkable work of Ornstein [44] which shows that Bernoulli shifts from a measure theoretical point of view can be classified by the entropy, a concept introduced earlier by Kolmogorov. But these results lead us far afield, and, in fact, Bernoulli shifts in the measure theoretical sense have, to my knowledge, not occurred in Celestial Mechanics.

I want to express my thanks to C. Conley, whose ideas and suggestions have been freely incorporated in the text, especially in Chapter VI. I also am indebted to him, and E. Zehnder, for reading the manuscript and providing fruitful criticism.

# CHAPTER II

# STABILITY PROBLEMS

## 1. *A model problem in the complex*

### a) *A stability theorem*

We illustrate the small divisor difficulty with a model problem which belongs to the theory of several complex variables. Consider a system of differential equations

$$\frac{dz}{dt} = f(z) = Az + \dots \tag{1.1}$$

where $z = (z_1, \dots, z_n)$ is a complex $n$-vector and $f(z)$ is also an $n$-vector whose components are holomorphic near $z = 0$. Moreover, we assume that $f(0) = 0$, so that $z = 0$ is an equilibrium solution whose stability is to be invectigated; $A$ denotes the matrix of the first derivatives of $f$ at $z = 0$.

We use the standard definitions of stability: The equilibrium solution $z = 0$ is said to be stable for $t > 0$ (future stability) if for every neighborhood $U$ of $0$ there exists a neighborhood $V$ with $0 \in V \subset U$ such that $z(0) \in V$ implies $z(t) \in U$ for $t > 0$. Similarly we define stability for $t < 0$ (past) and for all real $t$ $(-\infty < t < +\infty)$. In a sense, stability requires that the solutions depend continuously on its initial data, and while such statements are trivial for compact intervals, they lead to difficult problems for the half axis $t > 0$ or the whole axis.

It is a consequence of well-known results of A. Liapunov [45] that a necessary condition for future stability of $z = 0$ in (1.1) that the eigenvalues $\alpha_1, \alpha_2, \dots, \alpha_n$ of $A$ satisfy

$$\operatorname{Re} \alpha_k \leq 0 \,, \qquad k = 1, 2, \dots, n. \tag{1.2}$$

On the other hand, the condition $\operatorname{Re} \alpha_k < 0$ for $k = 1,2,\ldots,n$ is sufficient for future stability.

The above results are related to inequalities for so-called Liapunov functions, which decrease for increasing $t$. On the other hand, the stability problems in celestial mechanics, in which one has no friction, correspond here to the quest for stability for *all* real $t$. This is an entirely different problem and is related to *existence theorems* for some functional equations instead of *inequalities* in case of future stability. This is illustrated by the following surprising statement:

THEOREM 2.1 (Carathéodory-Cartan, 1932): *Necessary and sufficient conditions for the stability of the solution* $z = 0$ *of* (1.1) *for all real* $t$ *is that* i) $A$ *is diagonalizable with purely imaginary eigenvalues and* ii) *that there exists a holomorphic mapping*

$$z = u(\zeta) = \zeta + \ldots \tag{1.3}$$

*where* $\zeta$ *is a complex* $n$*-vector, taking* (1.1) *into the linear system*

$$\dot{\zeta} = A\zeta \ . \tag{1.4}$$

Clearly the condition i) means precisely that the linear system (1.4) is stable for all real $t$ and it is also evident that any system which can be transformed into (1.4) has a stable equilibrium solution since the concept of stability is invariant under coordinate transformations. Thus the sufficiency of the above conditions is obvious but the remarkable fact is that these conditions are also necessary. This is closely related to the compactness properties of bounded holomorphic families of functions. We will not present the proof of the theorem – which can be derived from the results of [46] – but mention only that the inverse mapping

$$\zeta = v(z) = z + \ldots$$

of (1.3) can be obtained by the formula

$$v(z) = \lim_{T\to\infty} \frac{1}{T} \int_0^T e^{-At}\psi(t,z)dt \tag{1.5}$$

where $\psi(t, z)$ is the solution of (1.1) with the initial value $\psi(0, z) = z$. In case of convergence of the above integral one has – on account of $\psi(t+s, z) = \psi(t, \psi(s, z))$ –

$$v(\psi(s, z)) = e^{As}v(z)$$

and differentiating with respect to $s$ at $s = 0$ we have

$$\sum_{k=1}^{n} f_k \frac{\partial v}{\partial z_k} = Av \tag{1.6}$$

where $f_k$ are the components of the vector $f$. Since $v = z + \dots$ is locally invertible this formula expresses that $\zeta = v(z)$ transforms (1.1) into (1.4). Thus the crux of the proof of Theorem 1.1 is the convergence of (1.5). In reference [46] the function $v(z)$ is constructed in a different way. It is shown that $e^{-At}\psi(t, z)$ for $z$ fixed in a sufficiently small neighborhood of the origin can be considered as a continuous function on the compact Abelian group generated by $e^{At}(-\infty < t < +\infty)$ and $v$ is obtained by averaging $e^{-At}\psi(t, z)$ over this Abelian group with respect to the Haar measure. However, it is easily shown that for any continuous function $\phi = \phi(g)$ on this group the average with respect to the Haar measure is given by

$$\lim_{T\to\infty} \frac{1}{T} \int_0^T \phi(e^{-At})dt$$

which leads to the formula (1.5).

b) *Implication of Theorem* 2.1

Instead of going further into the proof of this theorem we discuss its consequences: First of all it illustrates that the stability of the origin for all real $t$ – at least for this problem – is equivalent to an existence problem, namely to the existence for a solution $v(z) = z + \dots$ of the system of partial differential equations (1.6). Secondly, in case of stability for all real $t$ for (1.1) one can express the general solution of (1.1) near the origin in the form

$$z = u(e^{At}c)$$

where $c$ is a constant vector, $|c|$ sufficiently small. Writing $u(\zeta)$ as a power series and observing that the eigenvalues $\alpha_1, \alpha_2, \ldots, \alpha_n$ of $A$ are purely imaginary it follows that the solutions can be written in the form

$$z = \sum_j a_j e^{(j,\alpha)t}$$

where $a_j$ are complex numbers, $j = (j_1, \ldots, j_n)$ ranges over all vectors with nonnegative integer components $j_k$ and $(j, \alpha) = \sum_{k=1}^{n} j_k \alpha_k$. Since $u(z)$ is given by convergent power series it follows that

$$|a_j| \leq q^{|j|}$$

with a $q$ in $0 < q < 1$ and $|j| = j_1 + j_2 + \ldots + j_n \geq 1$. Therefore in case of stability all solutions of (1.1) near the origin are quasi-periodic functions with at most $n$ linearly independent frequencies. In fact, the whole flow is determined by, and even analytically equivalent to, the linearized system $\dot{z} = Az$, provided we have stability for all real $t$.

c) *Small divisors*

Although Theorem 2.1 provides a necessary and sufficient condition for stability it is quite useless for deciding about stability of a given system. It just shows that this question is equally difficult as that for the existence of a transformation (1.3) linearizing (1.1). To find a condition for the existence of $u$ we write it in the form

$$u = \zeta + \sum_{|j| \geq 2} u_j \zeta^j \, , \quad \zeta^j = \prod_{k=1}^{n} \zeta_k^{\,j_k} \tag{1.7}$$

and compare coefficients in the defining relation

$$(u_\zeta, A\zeta) = f(u(\zeta)) \; .$$

We may assume that A is a diagonal matrix, with the diagonal elements $\alpha_1, \alpha_2, \ldots, \alpha_n$. Then we obtain

$$((\alpha, j)I - A)u_j \ , \qquad (|j| \geq 2)$$

in terms of $u_\ell$ with $|\ell| < |j|$ from which we can determine the $u_j$ recursively and uniquely if the matrices $(\alpha, j)I - A$ are nonsingular, i.e., if

$$(\alpha, j) \neq \alpha_k \tag{1.8}$$

for all $k = 1, 2, \ldots, n$ and for integer vectors $j = (j_1, j_2, \ldots, j_n)$ with $j_\nu \geq 0$, $|j| \geq 2$. Thus under the conditions (1.8) the expansion for u of the form (1.7) is uniquely determined and it remains to investigate the convergence of these series to ensure stability.

First we show that violation of (1.8) generally leads to instability.

*Example*: Let $n = 2$ and $p, q$ be two integers, not both zero, satisfying $p \geq -1$, $q \geq 0$ and let

$$\alpha_1 p + \alpha_2 q = 0 \ ,$$

which means that (1.8) is violated for $j_1 = p+1$, $j_2 = q$. In this case the system

$$\dot{z}_1 = (\alpha_1 + z_1^p z_2^q) z_1$$

$$\dot{z}_2 = \alpha_2 z_2$$

has solutions satisfying $z_1^p z_2^q = (c - pt)^{-1}$ which implies the unstable character of the origin. It is obvious how to generalize this example to n dimensions.

We may assume that all $\alpha_1, \alpha_2, \ldots, \alpha_n$ are purely imaginary and set $\alpha_k = i\omega_k, \omega = (\omega_1, \omega_2, \ldots, \omega_n) \in R^n$. Let $\Omega$ denote the set of $\omega \in R^n$ for which (1.8) holds, and let $\tilde{\Omega} = R^n - \Omega$ be its complement. Furthermore let $Q_+$ denote the positive quadrant of those $\omega$ for which $\omega_k > 0 (k = 1, 2, \ldots, n)$, and similarly $Q_- = \{\omega \in R^n, \omega_1 < 0, \ldots, \omega_n < 0\}$. It is important to observe that $\tilde{\Omega}$ is dense in $R^n - Q_+ - Q_-$ while $Q_+ \cap \Omega$ and

$Q_- \cap \Omega$ are open sets. For example, for $n = 2$, $\tilde{\Omega} \cap (R^n - Q_+ - Q_-)$ consists of all rays through the origin with negative rational slope: $\omega_2/\omega_1 = -p/q$, while $\tilde{\Omega} \cap Q_+$ consists of the half rays $\omega_2/\omega_1 = 1/q$, $\omega_2/\omega_1 = q$, $q = 1, 2, \ldots$ only which are discrete (see Fig. 2).

It remains to discuss the convergence of the series u for $a_k = i\omega_k$, $\omega \in \Omega$. One has to distinguish again between $\omega \in \Omega \cap Q_+$, $\omega \in \Omega \cap Q_-$ and $\omega \in \Omega \cap (R^n - Q_+ - Q_-)$. In the first case, for a fixed $a$, the expressions $(a, j) - a_k$ are bounded away from zero, while in the second

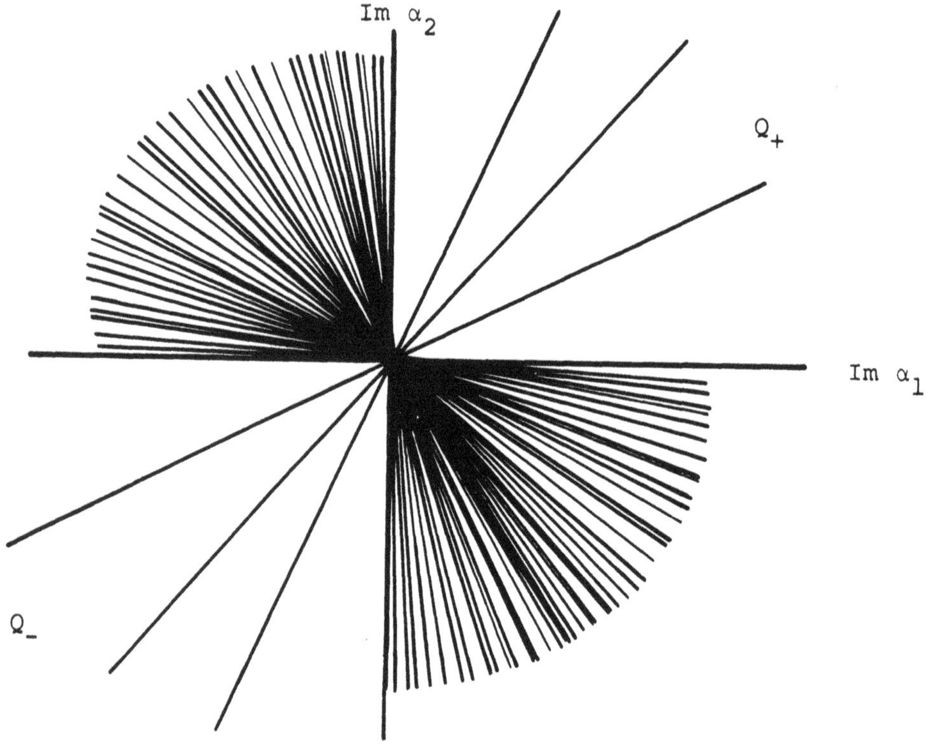


Fig. 2

they have zero as a cluster point. These numbers which enter in the denominator of $u_j$ in a complicated manner, are the so-called "small divisors" for this problem. Since in the quadrants $Q_+, Q_-$ these expressions are actually bounded away from zero, it is not difficult to verify that the series (1.7) for $u$ converge in a neighborhood of the origin. This implies that for $\omega \in \Omega \cap Q_+$ or $\omega \in \Omega \cap Q_-$ we have stability.

For $\omega \in \Omega \cap (R^n - Q_+ - Q_-)$, however, the situation is more delicate. The set of $\omega$ for which one has unstable systems is dense in $R^n - Q_+ - Q_-$ as we remarked already. On the other hand, the set $\Omega_s$ of $\omega$ for which $z = 0$ is stable for any system (1.1) with eigenvalues $a_k = i\omega_k$ is also dense in $R^n$. This is a consequence of the following

THEOREM 2.2 (Siegel [47]). *If the* $a_1, a_2, \ldots, a_n$ *satisfy for some positive constants* $c$, $\tau$ *the infinitely many inequalities*

$$|(j, a) - a_k| \geq c|j|^{-\tau} \tag{1.9}$$

*for all integer vectors* $j$ *with* $j_k \geq 0$, $|j| \geq 2$ *then the series* $u(\zeta) = \zeta + \ldots$ *transforming* (1.1) *into the linear system* (1.4) *converges in some neighborhood of the origin.*

If we take again $a_k = i\omega_k$ the set of $\omega$ satisfying (1.9) is dense in $R^n$; in fact, it is well known that the complement of this set is of Lebesgue measure zero in $R^n$, so that also $R^n - \Omega_s$ is of measure zero.

d) *Summary and interpretation*

The above results about the holomorphic systems (1.1) have the following surprising, and disturbing implications:

i) The origin is stable (for all real $t$) if and only if the system can be transformed into the linear one which is stable. Hence it is natural to express the conditions on stability in terms of the eigenvalues $a_k = i\omega_k$ of the matrix $A$.

ii) The set $\Omega_s$ of $\omega \in R^n$ for which all systems (1.1) with eigenvalues $a_k = i\omega_k$ are stable is dense in $R^n$. Moreover, almost all $\omega \in R^n$ belong to $\Omega_s$.

iii) The complement $\tilde{\Omega}_s = R^n - \Omega_s$ is dense in $R^n - Q_+ - Q_-$.

Thus, if $\omega \notin Q_+ \cup Q_-$ small perturbations of the system may make it unstable. In fact, it is easy to construct $f_1(z)$, $f_2(z)$, holomorphic near $z = (z_1, z_2) = 0$ such that the system

$$\dot{z}_1 = \alpha_1(z_1 + f_1(z))$$

$$\dot{z}_2 = \alpha_2(z_2 + f_2(z))$$

considered as dependent on the purely imaginary parameters $\alpha_1, \alpha_2$, is unstable for all negative rational values of $\alpha_2/\alpha_1$. On the other hand, for almost all values of $\alpha_2/\alpha_1$ one has stability.

Such a situation is clearly unacceptable for physical applications, since there all parameter values are known only with a certain degree of accuracy. The analytic reason for the density of the unstable system lies in the presence of the small divisors, which occur similarly in systems of physical significance and so we are led to the paradoxical situation mentioned in the Introduction. We will see, however, that in those systems of physical applications the nonlinearities may have an essentially stabilizing effect which is not the case in our model problem. In other words, although the above model problem illustrates clearly the difficulty of the small divisors and their effect on the stability of the system, it has to be considered as unrealistic and even misleading: Problems of applications are as a rule real systems, and not complex analytic, and for real systems we will see that the nonlinearities play an essential role. Historically the above model problem was the first one for which the small divisor difficulty was overcome (see [47], [48]).

e) *Some related results*

Theorem 1.1 related the stability question to the problem to linearize a system near an equilibrium via a coordinate transformation. This latter problem is of interest in itself and was the topic of the thesis of H. Poincaré [49] for real analytic systems of differential equations, where the partial differential equation (1.6) occurred and also conditions of the type (1.8) were again of importance. In the real case one may try to avoid these unpleasant conditions (1.8) by replacing the analyticity of the coordinate transformations by less stringent smoothness assumptions. Results in this direction are due to S. Sternberg [50] and P. Hartman (see [51]). To quote one of these results we consider a system

$$\dot{x} = f(x) = Ax + g(x) \tag{1.10}$$

where $x \in R^n$ and $g(x)$ is an n-vector whose components are real continuously differentiable near $x = 0$ and $g(0) = 0$, $\partial g/\partial x = 0$ at $x = 0$. Furthermore, let $\alpha_1, \alpha_2, \ldots, \alpha_n$ be the eigenvalues of $A$.

THEOREM 2.3 (P. Hartman). *If in* (1.10)

$$\operatorname{Re} \alpha_k \neq 0 \ , \qquad (k = 1, 2, \ldots, n) \tag{1.11}$$

*Then there exists a homeomorphism* $x = u(\xi)$ *of a neighborhood of* $\xi = 0$ *mapping the solutions of* (1.10) *into those of the linear system*

$$\dot{\xi} = A\xi \ .$$

For a proof see [52].

Thus the conditions (1.8) are replaced by (1.11) unfortunately excluding just the situation of purely imaginary eigenvalues which interests us here. It is remarkable, incidentally, that the homeomorphism $x = u(\xi)$ need not be differentiable, even if $g(x)$ is $C^\infty$ or real analytic. To obtain higher smoothness of $u(\xi)$ one has to impose at least some of the conditions (1.8) (see [51]).

The above theorem by Hartman applies, in particular, to the case $\operatorname{Re} \alpha_k < 0$ of future stability. One may expect, at least in this case to obtain the inverse mapping of $x = u(\xi)$ by

$$\lim_{t \to \infty} e^{-At} \psi(t, x)$$

analogously to (1.5), where $\psi(t,x)$ represents the solution of our system with $\psi(0,x) = x$. However, simple examples show that this limit need not exist (see, [53], [54]).

To give an interpretation of the quantities $(j,\alpha) - \alpha_k$ entering (1.8) Sternberg [55] considered the Lie algebra of vector field

$$F = \sum_{k=1}^{n} f_k(z) \frac{\partial}{\partial z_k}$$

where $f_k$ are formal power series without constant terms. The elements with $f_k(z) = \beta_k z_k$ form a commutative subalgebra and if we set

$$L = \sum_{k=1}^{n} \alpha_k z_k \frac{\partial}{\partial z_k}$$

then the operator

$$F \to [L,F] = ad_L F$$

has precisely $(j,\alpha) - \alpha_k$ eigenvalues, $z^j = \prod_{k=1}^{n} z_k^{j_k}$ being the corresponding eigenfunctions. Thus the condition (1.8) for $|j| \geq 1$ assures that the above commutative algebra is maximal, and L plays the role of a regular element in this Lie algebra. All these concepts have been developed for finite dimensional Lie algebras, while in the infinite dimensional case, needed here, they have only a formal meaning.

## 2. *Normal forms for Hamiltonian and reversible systems*

### a) *Theorem of Birkhoff*

After the consideration of the rather artificial model problem we turn to more realistic equations. The equations of mechanics can be obtained from variational principles, which lead along traditional lines to Hamiltonian systems of differential equations

$$\dot{x}_k = H_{y_k}, \quad \dot{y}_k = -H_{x_k}, \qquad k = 1,2,\ldots,n \tag{2.1}$$

where H is a real function of the 2n real variables $x = (x_1, \ldots, x_n)$, $y = (y_1, \ldots, y_n)$. It is well known that the form of these equations is preserved under the group of canonical coordinate transformations, that is, under coordinate transformations

$$x = u(\xi, \eta) \ , \quad y = v(\xi, \eta) \tag{2.2}$$

where u, v are vector functions of $\xi = (\xi_1, \ldots, \xi_n)$, $\eta = (\eta_1, \ldots, \eta_n)$ for which the identity

$$\sum_{k=1}^{n} dy_k \wedge dx_k = \sum_{k=1}^{n} d\eta_k \wedge d\xi_k \tag{2.3}$$

holds. Therefore this differential form, which is invariant under such coordinate transformations will play an important role in the following.

For simplicity of the exposition we consider the above system near an equilibrium solution, which we take as the origin of the coordinate system. Therefore we may assume that H vanishes with its first derivatives at the origin. We ask for a canonical coordinate transformation (2.2) which leads to a simplified system, to a so-called normal form. This will not any more be a linear system, and in this lies the main difference from the previous section. Otherwise the problem is entirely analogous to that of Section 1, except that the group of holomorphic transformations is replaced by that of canonical transformations.

We consider the linear part of the system (2.1), determined by the second derivatives of H. The matrix of this linear system has the property that if $\alpha$ is an eigenvalue, so are $\bar{\alpha}$, $-\alpha$, $-\bar{\alpha}$ eigenvalues. The first follows from the real character, the second from the Hamiltonian nature of the system. Therefore we conclude, from theorems of Liapunov, alluded to in Section 1 that one can have future stability only if all eigenvalues lie in $\text{Re}\,\alpha \leq 0$, and hence only if all eigenvalues are purely imaginary. This is a reflection of the fact that (2.1) describes nondissipative motions.

Therefore we will assume that all eigenvalues are purely imaginary and distinct, and we label them such that $\alpha_1, \alpha_2, \ldots, \alpha_n$, $-\alpha_1, -\alpha_2, \ldots, \alpha_n$

are all the eigenvalues. It is easily shown that there exists a real canonical transformation that the quadratic terms in the Taylor expansion of H have the form

$$\sum_{k=1}^{n} \frac{i\alpha_k}{2}(x_k^2 + y_k^2)\ , \quad \alpha_k = -\alpha_k\ . \tag{2.4}$$

Thus if H would be just this quadratic function the motion would be described by n uncoupled harmonic oscillations. The problem, now, is to study the effect of the nonlinearities.

In order to avoid any convergence questions at first we consider H, as well as the transformations (2.2) as formal power series. The composition of such transformations, as formal power series is well defined if they have no constant terms. If we require, moreover, that the Jacobian at the origin $\neq 0$, they form a group. Combining the 2n components of x, y to one vector $z = (x, y)$ and, similarly, $\zeta = (\xi, \eta)$ we write (2.2) in the form

$$z = w(\zeta)$$

where w consists of 2n formal power series in $\zeta_1, \ldots, \zeta_{2n}$ without constant terms satisfying the formal identities (2.3). By standard results of transformation theory the transformed system is given by the Hamiltonian

$$\Gamma(\zeta) = H(w(\zeta))$$

obtained by transforming H. Now we ask for a transformation of this type simplifying the Hamiltonian, i.e., for a normal form.

THEOREM 2.4 (G. D. Birkhoff [4]). *If* H *is a formal, real power series without constant and linear terms and quadratic terms of the form* (2.4) *where* $\alpha_1, \alpha_2, \ldots, \alpha_n$ *are purely imaginary and independent over the rationals then there exists a formal real canonical transformation* $z = w(\zeta)$ *such that*

$$H(w(\zeta)) = \Gamma(\xi_1^2 + \eta_1^2, \ldots, \xi_n^2 + \eta_n^2)$$

*is a power series in* $\xi_k^2 + \eta_k^2$.

For the proof of this theorem we refer to [8]. If the above series H, w hence $\Gamma$ would converge we could integrate the resulting Hamiltonian system readily: Denoting the derivative of $\Gamma$ with respect to the argument $\xi_k^2 + \eta_k^2$ by $\Gamma_k$ we have

$$\frac{\partial\Gamma}{\partial\xi_k} = 2\xi_k\Gamma_k\,, \quad \frac{\partial\Gamma}{\partial\eta_k} = 2\eta_k\Gamma_k$$

and therefore the transformed system would take the form

$$\dot{\xi}_k = 2\eta_k\Gamma_k\,, \quad \dot{\eta}_k = -2\xi_k\Gamma_k\,. \tag{2.5}$$

This implies that $(\xi_k^2 + \eta_k^2)^{\cdot} = 0$ and since $\Gamma$ and $\Gamma_k$ depend only on $\xi_1^2 + \eta_1^2, \ldots, \xi_n^2 + \eta_n^2$ we have also $\frac{d}{dt}\Gamma_k = 0$, and $\Gamma_k$ can be considered as real constants. Thus the integration of (2.5) yields

$$\xi_k + i\eta_k = e^{-2it\Gamma_k}(\xi_k(0) + i\eta_k(0))\,,$$

where the "frequencies" $2\Gamma_k$ depend on the "amplitudes" $\xi_\ell^2(0) + \eta_\ell^2(0)$ $(\ell = 1, \ldots, n)$. Inserting this result into (2.2) we obtain x, y as quasi-periodic functions, and also stability would be insured.

b) *Convergence question*

The above discussion was crucially dependent on the convergence of the series entering into the transformation $z = w(\zeta)$. A further difficulty of Theorem 2.4 lies in the assumption that the $\alpha_k$ are rationally independent, since the set of real n-vectors which are rationally dependent are dense in $R^n$, even though they are of measure zero. This is clearly a similar situation as for the holomorphic system of Section 1, and one may expect a similar answer to the question of convergence. However, according to a result of Siegel the divergence of the above series represents the general case! We make this precise for $n = 2$ only and assume that the Hamiltonian is of the form

$$H = F + G$$

where $F$ is a fixed real polynomial of $p_1 = x_1^2 + y_1^2$ and $p_2 = x_2^2 + y_2^2$ of degree $s \geq 2$ starting with (2.4) while

$$G = \sum_{|\nu|+|\mu|\geq 2s} g_{\nu\mu} x^\nu y^\mu ,$$

is a real analytic function where $\nu = (\nu_1, \nu_2)$, $\mu = (\mu_1, \mu_2)$, $x^\nu = x_1^{\nu_1} x_2^{\nu_2}$. Thus $F$ is a fixed Hamiltonian in normal form for which we assume that $a_1/a_2$ is irrational and

$$a_1 F_{p_2} - a_2 F_{p_1} \not\equiv 0 .$$

The second part is considered variable and we may assume that the coefficients are restricted to

$$|g_{\nu\mu}| < 1 . \tag{2.6}$$

THEOREM 2.5 (Siegel [56]). *There exist infinitely many power series* $\Phi_\ell (\ell = 1,2,\ldots)$ *in the coefficients* $g_{\nu\mu}$, *absolutely convergent in* (2.6) *such that the transformation into normal form for* $H$ *is convergent only if* $\Phi_\ell = 0$ *for infinitely many* $\ell$. *The* $\Phi_\ell$ *are analytically independent.*

This theorem shows clearly that the divergence is the general case. Moreover, for this it is quite irrelevant whether the irrational number $a_1/a_2$ can be well approximated by rationals or not, in contrast to Theorem 2.2.

This result seems to make the normal form of Theorem 2.4 quite useless. Of course, by truncating the above series one obtains at least a good approximation of the solutions by trigonometrical sums. Beyond this, we will see later that the above normal form will prove very useful for deciding the stability question.*

c) *Reversible systems*

We wish to discuss a normal form for another class of differential equations, the reversible systems, to which the n-body problem also belongs. The following definition is motivated by the example of a second order system

* For a general study of normal forms for systems of differential equation see Brjuno [57].

(2.7) $$\ddot{q}_k = Q_k(q) , \qquad (k = 1,\dots, m)$$

which has the property that with $q(t) = (q_1(t),\dots, q_m(t))$ also the reversed function $q(-t)$ is a solution.

More generally, we consider a first order system

$$\dot{x} = f(x)$$

where $x$ and $f$ are n-vectors. We assume there exists a linear reflection $R$ of $R^n$, so that

(2.8) $$R^2 = I$$

and

(2.9) $$f(Rx) = -Rf(x) .$$

Thus with $x(t)$ also $Rx(-t)$ is a solution. Under these circumstances we call the above system reversible. Taking $n = 2m$,

$$x = \begin{pmatrix} q \\ \dot{q} \end{pmatrix}, \qquad Rx = \begin{pmatrix} q \\ -\dot{q} \end{pmatrix}$$

it is clear that (2.7) corresponds to a reversible first order system.

We discuss a reversible system near an equilibrium, say $x = 0$, and begin with a normal form for the linearized system

(2.10) $$\dot{x} = Ax$$

under the assumption of stability. According to (2.8) $R$ possesses only the eigenvalues $+1$ and $-1$, and let $d_+$, $d_-$ denote the dimensions of the corresponding eigenspaces, so that $d_+ + d_- = n$. The condition (2.9) implies

(2.11) $$AR = -RA .$$

One shows easily that

$$\operatorname{rank}(A) \le 2 \min(d_+, d_-)$$

and for A to be nonsingular we have to assume* that n is even and

$$d_+ = d_- = \frac{n}{2} .$$

From (2.11) we read off that with $\alpha$ also $-\alpha$ is an eigenvalue of A, so that future stability of the linear system can take place only if all eigenvalues are purely imaginary. We assume that all eigenvalues are distinct. With some elementary arguments one verifies that under the above assumptions one can find coordinates $u_k, v_k (k = 1,\ldots,m)$ such that (2.10) takes the form

$$\begin{aligned} \dot{u}_k &= -\beta_k v_k \\ \dot{v}_k &= +\beta_k u_k \end{aligned} \qquad (k = 1,\ldots,m)$$

with $\alpha_k = i\beta_k$ and

$$R : (u,v) \to (u,-v) .$$

Introducing complex coordinates $z_k = u_k + iv_k$ we can write the linear system in the form

$$\dot{z}_k = \alpha_k z_k \quad \text{and} \quad R : z \to \bar{z} . \tag{2.12}$$

With this simplification we consider reversible systems of the form

$$\dot{z} = f(z,\bar{z}) \tag{2.13}$$

where f is a vector valued power series in $z_k, \bar{z}_k$ satisfying, according to (2.9),

$$\bar{f}(\bar{z},z) = -\overline{f(z,\bar{z})} \tag{2.14}$$

and according to (2.12)

$$\left.\frac{\partial f}{\partial z}\right|_{z=0} = \operatorname{diag}(\alpha_1,\alpha_2,\ldots,\alpha_n); \quad \bar{\alpha}_k = -\alpha_k . \tag{2.15}$$

---

* Actually the following results can be extended to the case $d_+ \neq d_-$, but have a somewhat more complicated form.

We observe that the class of reversible differential equations is transformed into itself under the pseudo-group of coordinate transformations which commute with R. This transformation group replaces that of the canonical transformations above, and again we ask for a normal form for (2.13).

To avoid convergence questions we consider at first the components of $f(z,\bar{z})$ as formal power series and admit coordinate transformations

$$z = u(\zeta,\bar{\zeta}) = \zeta + \dots \tag{2.16}$$

whose components are formal power series; in order that these formal transformations commute with R we require

$$u(\bar{\zeta},\zeta) = \overline{u(\zeta,\bar{\zeta})} \, . \tag{2.17}$$

Analogously to Theorem 2.4 one has

THEOREM 2.6. *If the system* (2.13) *satisfies* (2.14), (2.15) *and if the eigenvalues* $\alpha_k$ *in* (2.15) *are rationally independent then there exists a formal coordinate transformation* (2.16) *satisfying* (2.17) *such that* (2.13) *is transformed into*

$$\dot{\zeta}_k = \zeta_k \Phi_k \tag{2.18}$$

*where* $\Phi_k = \alpha_k + \dots$ *are formal power series in the* m *products* $\zeta_1\bar{\zeta}_1, \dots, \zeta_m\bar{\zeta}_m$ *with purely imaginary coefficients.*

The proof proceeds by comparison of coefficients and will be omitted. It is important that the $\Phi_k$ are purely imaginary which is a consequence of (2.14). It implies that, at least, if we assume all series to be convergent, that for any solution $\zeta_k\bar{\zeta}_k$ is independent of t, hence also $\dot{\Phi}_\ell = 0$, so that the solutions are given by

$$\zeta_k(t) = e^{t\Phi_k}\,\zeta_k(0)$$

where the arguments of $\Phi_k$ are $|\zeta_1(0)|^2, \dots, |\zeta_m(0)|^2$. Thus, since $\Phi_k$

are purely imaginary we are again led to quasi-periodic solutions, aside from the question of convergence. One has to expect that for this problem divergence is the typical case, like in Theorem 2.5, although, to my knowledge such a result has not been established.

## 3. *Invariant manifolds*

### a) *Motivation*

In the two preceding sections we found normal forms for systems of differential equations, in the case of holomorphic systems and Hamiltonian systems. In both cases we were led to the requirement of rational independence of the eigenvalues, giving rise to the small divisor difficulty, i.e., to the problem of convergence. We saw that in the holomorphic case convergence took place in general while for the Hamiltonian case divergence was generic. These statements exhibit clearly the difficulties mentioned already in Chapter I, and on account of the requirement of rational independence of the eigenvalues the set of admitted systems is not open in the class considered, and we come to the question of how to overcome these obstacles which contradict the physical intuition.

The basic idea is to give up the description of *all* solutions near the equilibrium, but instead to study the solution on some invariant submanifolds only. For this we recall the nonlinear normal form (2.5) of Theorem 2.4. In case of convergence of these series the system (2.5) describes oscillation with the frequencies $2\Gamma_k = i\alpha_k + \dots$ . Thus if we restrict attention to a torus

$$\xi_k^2 + \eta_k^2 = c_k \,, \qquad\qquad c = (c_1, c_2, \dots, c_n)$$

with sufficiently small positive $c_k$ these frequencies $2\Gamma_k(c)$ will, in general, differ from $i\alpha_k$ and may well be rationally independent, even if the $\alpha_k$ are not, provided $c$ is appropriately chosen. This requires that the frequencies $2\Gamma_k(c)$ change as the $c_1, \dots, c_n$ vary, i.e., that the frequencies do change with the amplitudes, which is a typical phenomenon for nonlinear problems. In this respect the problems of Section 1 differed from those of Section 2 as the former gave rise only to linear normal forms. The nonlinear terms in the normal form in Section 2 will be essential and will, in some sense, have a stabilizing effect.

b) *Subsystems*

If $f$ is a vector field on a manifold $M$ and $\phi$ a vector field on another manifold $N$ we call $\phi$ a subsystem of $f$ if i) there exists an embedding $w: N \to M$ of $N$ in $M$ taking $\phi$ into $f$, restricted to $w(N)$. In coordinates, if

(3.1) $$\dot{z} = f(z) \ , \quad \dot{\zeta} = \phi(\zeta)$$

describe the vector fields and

$$z = w(\zeta)$$

the embedding we have to satisfy the system of partial differential equations

$$\sum_k \frac{\partial w}{\partial \zeta_k} \phi_k = f(w) \ .$$

Geometrically, this means, $N$ is a submanifold consisting of solutions, i.e., $N$ is an "invariant manifold" under the flow defined by $f$, and $\phi$ describes the vector field $f$ restricted to $N$ in appropriate coordinates. Generally we will require $N$ to be a compact manifold.

As a trivial example we consider a periodic solution which gives rise to a subsystem. Writing a periodic solution of (3.1) in the form

$$z = p(\omega t)$$

where the positive number $\omega$ is chosen such that $2\pi$ is the smallest period of $p(\zeta)$; here $\zeta$ is a scalar variable. Since (3.1) is independent of $t$ also $p(\omega(t-t_0))$ is a solution and we can consider

$$z = p(\zeta)$$

an embedding of a circle parametrized by $\zeta \pmod{2\pi}$, and the corresponding subsystem is simply

$$\dot{\zeta} = \omega \ .$$

More generally, if

$$z = p(\zeta_1, \ldots, \zeta_s) \tag{3.2}$$

is an embedding of an s-dimensional torus, parametrized by $\zeta = (\zeta_1, \zeta_2, \ldots, \zeta_s) \pmod{2\pi}$ then the system

$$\dot{\zeta}_\sigma = \omega_\sigma , \qquad (\sigma = 1, 2, \ldots, s) \tag{3.3}$$

gives rise to a family of quasi-periodic solutions. We may assume that the $\omega_1, \ldots, \omega_s$ are rationally independent in which case they are called a frequency basis of the quasi-periodic solutions.

Geometrically (3.2) represents the embedding of a torus and (3.3) a flow on the torus. If the $\omega_1, \ldots, \omega_s$ are rationally independent each orbit of this flow is dense on the torus and the torus can be viewed as the closure of any orbit lying on it.

For example, in case of convergence of the normal form (2.5) we can view $\xi_k^2 + \eta_k^2 = c_k$ $(k = 1, 2, \ldots, n)$ with $c_k > 0$ as an invariant torus, and

$$\xi_k = \sqrt{c_k} \cos \zeta_k , \quad \eta_k = \sqrt{c_k} \sin \zeta_k$$

represents an embedding of the torus, and

$$\dot{\zeta} = -2\Gamma_k(c)$$

the corresponding subsystem.

In the following we will study the perturbation theory of such subsystems, and exhibit conditions under which such invariant submanifolds persist under perturbation of the vector field. We wish to emphasize that the perturbations of the vector fields have to be restricted to a subclass of vector fields, like Hamiltonian systems or reversible systems, and for general perturbation such a perturbation theory does not exist. This is a reflection of the fact, that dissipation of a system generally destroys quasi-periodic solution. Before describing such a perturbation theory we introduce the class of integrable Hamiltonian systems which represent the unperturbed systems.

c) *Integrable systems*

A Hamiltonian system (3.1) is called integrable if one can introduce coordinates $p = (p_1, \ldots, p_n)$, $q = (q_1, \ldots, q_n)$ via a canonical transformation

$$z = w(p, q) \tag{3.3}$$

such that $w$ has period $2\pi$ in the variables $q_1, q_2, \ldots, q_n$ and such that the Hamiltonian $H = H(p)$ describing this system is independent of $q$, so that the system takes the simple form

$$\dot{q}_k = H_{p_k}, \quad \dot{p}_k = -H_{q_k} = 0 . \tag{3.4}$$

Thus, $p_k$ are constant along each solution, i.e., the $p_k$ are "integrals" and the system (3.4) is easily solved by

$$q_k = H_{p_k}(c)t + q_k(0), \quad p_k = c_k$$

justifying the term "integrable" system.

One has to bear in mind that this definition depends on the choice of coordinates; the above canonical coordinates are referred to as "normal coordinates," or sometimes as action and angle variables. In the older history of this subject one thought of solving a dynamical problem as finding such normal coordinates, taking their existence for granted. The Hamiltonian-Jacobi theory was, in fact, designed for this purpose. To see this connection we write $z = (x, y)$, $x = (x_1, \ldots, x_n)$, $y = (y_1, \ldots, y_n)$ in (3.3) and

$$\begin{cases} y = u(p, q) \\ x = v(p, q) \end{cases} \tag{3.5}$$

and the canonical character means that

$$\sum_{k=1}^{n} dy_k \wedge dx_k = \sum_{k=1}^{n} dp_k \wedge dq_k .$$

There is a standard representation of such canonical transformation in terms of a generating function. Assuming that the second equation in (3.5) can be solved for q, in terms of the independent variables p, x, the equation (3.5) takes the form

$$q_k = \frac{\partial S}{\partial p_k}, \quad y_k = \frac{\partial S}{\partial x_k}$$

where S, the generating function, depends on p and x. Conversely, any such function S gives rise to a canonical transformation, provided one can express x, y in terms of p, q. If $G(x, y)$ is the Hamiltonian for (3.1) and $H(p, q)$ the Hamiltonian in the variables p, q one has

$$G(x, S_x) = H(p, S_p) .$$

In order that p, q are normal coordinates, H must be independent of q, i.e.,

$$G(x, S_x) = H(p) . \tag{3.6}$$

Thus, finding normal coordinates is closely related to finding a solution S of the above Hamiltonian-Jacobi equation, where S has to depend on n parameters $p_1, p_2, \ldots, p_n$ such that $\det S_{xp} \neq 0$. It is well known that such a solution of the Hamilton-Jacobi equation always exists locally. This simply reflects that locally, near a nonsingular point, all Hamiltonian vector fields are equivalent under a canonical coordinate change, and locally all Hamiltonian systems possess n linear independent integrals in involution. But the question of finding normal coordinates is a global one as they should describe the neighborhood of a torus. For this purpose we have to impose periodicity conditions on S which we do not formulate here since we come back to this point in d).

A number of problems were solved by this method, and Jacobi succeeded in applying this approach in a surprising way. He found, for example, that the geodesic flow on an ellipsoid with three different principal axes is integrable; in this case the normal coordinates are related to the curvature lines. Other examples are the planar motion of a mass point under the gravitational attraction of two fixed mass points, a problem going back to Euler. Although this latter problem seems very close to the restricted three-body problem one did not succeed in finding such normal coordinates for the latter problem. In fact, it turns out that this problem is not integrable, an assertion going back to Poincaré.

Another example of an integrable system is given by the normal form (2.5) of Theorem 2.4, *if it is convergent*. Indeed, if one sets

$$\xi_k + i\eta_k = \sqrt{2p_k}\, e^{iq_k}$$

in (2.5) then $p_k$, $q_k$ are normal coordinates. Theorem 2.5 can be interpreted as the assertion that, in general, Hamiltonian systems near an equilibrium are not integrable, i.e., such normal coordinates do not exist in general.

A simpler example of an integrable system is the two-body problem, and similarly the N-body problem I(1.1) in the limit case where $m_2 = m_3 = \dots = m_N = 0$. Thus the planetary problem can be viewed as a perturbation of an integrable system.

It is worth mentioning that even N. Bohr's quantum theory was based on the concept of action and angle variables [58], as Bohr's quantization restricted the value of the action variables to integer multiples of the Planck constant. For the hydrogen atom, which is based on the two-body problem, this gives rise to correct energy levels, but even for the non-integrable three-body problem this approach fails, and may at most have an asymptotic meaning.

d) *Perturbation of integrable systems*

Now we consider a small perturbation of an integrable system. Let $x_k$, $y_k (k = 1, 2, \dots, n)$ be the normal coordinates for the unperturbed system, given by a Hamiltonian $H_0(y)$ and assume that the perturbed system is given by

$$H(x, y, \mu) = H_0(y) + \mu H_1(x, y) + \dots \tag{3.7}$$

where $\mu$ is a small parameter and H is assumed to be a real analytic function, the 2n+1 variables. Moreover, H is required to have the period $2\pi$ in $x_1, x_2, \ldots, x_n$. Thus, for $\mu = 0$ the 2n-dimensional phase space is foliated into an n-parameter family of invariant tori $y_k = c_k (k = 1, \ldots, n)$ on which the flow is given by

$$\dot{x}_k = \frac{\partial H_0(c)}{\partial y_k} .$$

What happens to this foliation for small values of $\mu$? It turns out that those tori for which the frequencies

$$\omega_k = \frac{\partial H_0(c)}{\partial y_k} \tag{3.8}$$

are not only rationally independent, but satisfy with some positive constants $\gamma$, $\tau$ the inequalities

$$\left| \sum_{k=1}^{n} j_k \omega_k \right| \geq \gamma |j|^{-\tau} \tag{3.9}$$

for all integers $j_k$ with $|j| = \sum |j_k| \geq 1$, persist under perturbation for small values of $\mu$. This is the content of the following

THEOREM 2.7 (Kolmogorov-Arnold [59], [6]). *Let* Y *be an open set in* $R^n$ *and let* $H(x, y, \mu)$ *be a real analytic function of* x, y, $\mu$ *for all real* $x, y \in Y$ *and near* $\mu = 0$. *Moreover,* H *is assumed to have the period* $2\pi$ *in* $x_1, x_2, \ldots, x_n$ *and* $H_0 = H(x, y, 0)$ *to be independent of* x. *Let* c *be chosen so that the frequencies* (3.8) *satisfy* (3.9) *and that the Hessian*

$$\det \left( \frac{\partial^2 H_0}{\partial y_k \, \partial y_\ell} \right) \tag{3.10}$$

*does not vanish at* $y = c$.

*Then for sufficiently small* $|\mu|$ *there exists an invariant torus*

$$\begin{cases} x = \theta + u(\theta, \mu) \\ y = c + v(\theta, \mu) \end{cases} \tag{3.11}$$

*where* $u(\theta,\mu)$, $v(\theta,\mu)$ *are real analytic functions of* $\mu$ *and* $\theta = (\theta_1,\ldots,\theta_n)$, *have period* $2\pi$ *in* $\theta_1,\theta_2,\ldots,\theta_n$ *and vanish for* $\mu = 0$. *Moreover, the flow on this torus is given by*

$$\dot{\theta}_k = \omega_k\,, \qquad (k = 1,\ldots, n). \tag{3.12}$$

*This equation defines a subsystem of the system given by* (3.7) *and* (3.11) *defines the embedding of the torus.*

One may expect that such continuation holds for all tori of the unperturbed family. This, however, is not the case and it was known to Poincaré already that those tori $y = c$ for which the $\omega_k$ are integer multiples of one positive number and hence consist of periodic solutions, break up under small perturbations, and, in general, only a finite number of periodic solutions will persist for small values of $|\mu|$. Thus a condition of the type (3.9), excluding rationally dependent frequencies is quite essential. Of course, it appears unlikely that one can maintain such a relation under perturbation. Therefore, it is important to observe that the above theorem asserts that the frequencies $\omega_k$ are kept independent of $\mu$, and so all relations (3.9) are indeed preserved. That this is possible is a consequence of the condition (3.10).

Actually, more than asserted in Theorem 2.7 can be proven. It turns out that the differential form $\sum_{k=1}^{n} dy_k \wedge dx_k$ vanishes identically on the tori (3.11), and one calls manifolds with this property and of maximal dimension Lagrange manifolds. If we represent this torus in the explicit form

$$y_k = g_k(x,\mu)$$

it follows that

$$\sum_{k,\ell=1}^{n} g_{kx_\ell}\, dx_k \wedge dx_\ell = \sum_{k<\ell} (g_{kx_\ell} - g_{\ell x_k})\, dx_k\, dx_\ell = 0\,.$$

Hence, there exists a function $S(x,\mu)$ such that $g_k = S_{x_k}$ and the torus is given by

$$y_k = S_{x_k}(x, \mu) \ , \tag{3.13}$$

where the gradient of $S$ has period $2\pi$ in $x_1, \dots, x_n$.

The fact, that (3.13) represents an invariant torus is equivalent to

$$\dot{y}_k = \sum_\ell S_{x_k x_\ell} \dot{x}_\ell$$

or the identity

$$-\frac{\partial H}{\partial x_k} = \sum S_{x_k x_\ell} \frac{\partial H}{\partial y_\ell} \ ,$$

or to the statement that

$$H(x, S_x, \mu) = c(\mu)$$

is independent of $x$. Thus the above theorem asserts, indeed, the solvability of the Hamilton-Jacobi equation with a function $S$ whose gradient has period $2\pi$ in $x_1, x_2, \dots, x_n$ for small values of $|\mu|$. We discussed the solution of the Hamilton-Jacobi equation (3.6) in dependence of $n$ parameters $p_1, \dots, p_n$ for transforming the perturbed system into an integrable system and saw that this was generally not possible. But for some values of $p$, which correspond to frequencies $\omega$ satisfying (3.9) such a solution does exist and gives rise to an invariant torus (3.11). This answers precisely the question raised by Weierstrass when he doubted that Poincaré's theorem on the nonexistence of integrals conclusively implies the nonexistence of quasi-periodic solutions (see discussion in Chapter I).

As another remark to the fundamental Theorem 2.7 we point out that it provides not only the existence of a single torus for each small value of $\mu$, but gives a set of such tori corresponding to different choices of $\omega$. Since the complement of the set of $\omega$ satisfying (3.9) with some $\gamma$, $\tau$ of measure zero, one might expect that the tori so found leave out a set of measure zero only. This is, however, not the case since the tori exist only for $|\mu| \le \mu_0(\gamma)$ where $\mu_0$ depends on $\gamma$. However, the following assertion can be proven (see [59]):

We considered the above system in the domain $T^n \times Y$ where $T^n$ denotes the torus given by $(x, y)$ with $y = \text{const}$. We may assume that $Y$ has a compact closure, otherwise we consider such a subset. Now let $\varepsilon$ be a given positive number, then there exists a $\mu_1 = \mu_1(\varepsilon) > 0$ such that for $|\mu| < \mu_1$ the $\Sigma = \Sigma(\mu) \subset T^n \times Y$ covered by the invariant tori satisfies

$$m((T^n \times Y) - \Sigma) < \varepsilon m(T^n \times Y) ,$$

where $m$ denotes the Lebesgue measure in $T^n \times Y$. Thus the majority of the solutions in this set are quasi-periodic.

Although this theorem is not directly applicable to the equilibrium problem of Section 2 the ideas can be extended to give the following stability result: Using the notation of Section 2 we assume that

$$\sum_{k=1}^{n} j_k \alpha_k \neq 0 \quad \text{for} \quad 0 < |j| \leq 4 \tag{3.14}$$

for all integers $j_k$ with the above restriction. Then the normal form of Theorem 2.4 can be achieved at least to terms of order 4 and the transformed Hamiltonian $\Gamma = \Gamma(\xi, \eta)$ has the form

$$\Gamma = \sum_{k=1}^{n} i\alpha_k p_k + \sum \beta_{k\ell} p_k p_\ell + O_5$$

where

$$p_k = \frac{1}{2}(\xi_k^2 + \eta_k^2)$$

and $O_5$ is a power series containing terms of order $\geq 5$ only.

THEOREM 2.8. *Let* $H(x, y)$ *be a real analytic at the origin, whose first derivatives vanish there. If then* (3.14) *holds, and*

$$\det(\beta_{k\ell}) \neq 0 \tag{3.15}$$

*then for any* $\varepsilon > 0$ *there exists a neighborhood* $U = U(\varepsilon)$ *of the origin containing a set* $\Sigma = \Sigma(\delta)$ *of quasi-periodic solutions such that*

$$m(U - \Sigma) < \varepsilon m(U) .$$

Thus, under the finitely many conditions (3.14), (3.15) one has a weakened type of stability. Aside from a small proportion all solutions of this neighborhood U remain in it. Examples show that the excluded set of escaping solutions may not be empty.

The important point of this theorem, however, is that this weakened type of stability requires only finitely many conditions, and not the rational independence of the eigenvalues. The condition (3.15) restricts the nonlinear terms and effectively allows us to drop condition (3.14) for $|j| > 4$. Moreover, a flow possessing such an equilibrium solution is not ergodic, as the set $\Sigma$ is an invariant set of positive measure.

The above can be generalized to the case where not the fourth order terms in the Hamiltonian but those of some even order are nondegenerate. Also it is possible to free ourselves from the class of real analytic functions and require that H belongs to $C^\ell$ where $\ell$ is large enough. In [60] it was shown that $\ell > 2n + 2$ suffices, or even that the derivatives up to order $2n + 2$ are Hölder continuous, is enough.

e) *Reversible systems*

There is an analogous statement to Theorem 2.7 for the class of reversible systems. Since the proof for this case is technically somewhat simpler we will present it in Chapter V, instead for the Hamiltonian case. But the ideas are, in fact, the same.

We consider a system

$$\dot{x} = f(x, y, \mu)$$

$$\dot{y} = g(x, y, \mu)$$

where x, y are n-vectors, f, g real analytic functions, of period $2\pi$ in $x_1, \ldots, x_n$ and

$$f(x, y, 0) = F(y) \ , \quad g(x, y, 0) = G(y)$$

independent of x. The reversible character is expressed by

$$f(-x, y, \mu) = f(x, y, \mu) \ , \quad g(-x, y, \mu) = -g(x, y, \mu) \ .$$

In particular it follows that $G(y) = 0$, and the system for $\mu = 0$ admits $y = c$ as invariant tori with the flow

$$\dot{x} = F(c) \ .$$

By requiring that the Jacobian determinant of $F_y$ does not vanish we will assure that one can choose $c$ in such a way that frequencies $\omega_k = F_k(c)$ satisfy (3.9).

THEOREM 2.9. *Under the above conditions, if*

$$\det\left(\frac{\partial F_k}{\partial y_\ell}\right) \neq 0 \quad \text{at} \quad y = c$$

*there exists an invariant torus*

$$x = \theta + u(\theta, \mu)$$

$$y = c + v(\theta, \mu)$$

*where* $u(\theta,\mu)$, $v(\theta,\mu)$ *are real analytic, of period* $2\pi$ *in* $\theta_1, \theta_2, \ldots, \theta_n$ *and vanish for* $\mu = 0$, *with the flow*

$$\dot{\theta}_k = \omega_k$$

*on it. The reversible character implies, moreover,*

$$u(-\theta, \mu) = -u(\theta, \mu) \ , \qquad v(-\theta, \mu) = v(\theta, \mu) \ .$$

Also Theorem 2.8 has an analogue: If in the system (2.13) the purely imaginary eigenvalues $a_1, a_2, \ldots, a_n$ satisfy (3.14) then the normal form can be obtained up to terms of order 4 where, in the notation of the previous section,

$$\Phi_k(p) = a_k + i \sum_{\ell=1}^{n} \beta_{k\ell} p_\ell \ , \qquad p_k = |\zeta_k|^2$$

and $(\beta_{k\ell})$ is a real symmetric matrix. If this matrix is nonsingular one has the same weakened stability statement as in Theorem 2.8.

## 4. *Twist theorem*

### a) *Annulus mapping*

This section is devoted to some results on area-preserving mappings in the plane. We will discuss a number of theorems, quite analogous to those of the previous section, but the results of this section have the advantage to be more accessible to geometrical interpretation.

There is a well-known and close connection between flows and mappings. For example, if one is interested in the flow in an m-dimensional phase space near a periodic orbit one can associate with this flow a mapping of an $(m-1)$-dimensional neighborhood of a fixed point of this mapping. For this purpose one simply intersects the periodic orbit transversally with some manifold $\sigma$ of $(m-1)$-dimension and defines the mapping $\phi$ by following an orbit from a point of $\sigma$ to the next intersection with $\sigma$ for increasing $t$. This mapping has the point of intersection of $\sigma$ and the periodic orbit as a fixed point, and is clearly well defined near this fixed point.

For Hamiltonian systems of two degrees of freedom $(m = 4)$ on a fixed energy surface the same type of construction gives rise to a mapping of a planar region, preserving an area element (see [8]). Such mappings reflect most of the interesting features of the flow; for example, periodic orbits correspond to fixed points of some iterates of $\phi$, stability of a periodic orbit is equivalent to stability of the corresponding periodic point of $\phi$, etc.

Area-preserving mappings in the plane were studied extensively by Birkhoff [62] and we begin with an interesting fixed point theorem which was stated by Poincaré and later proven by Birkhoff [61]. This theorem deals with an area-preserving mapping defined and one-to-one in an annulus

$$A : a \leq r \leq b \, , \qquad 0 < a < b, \tag{4.1}$$

given by

$$r_1 = f(r,\theta) \, , \quad \theta_1 = \theta + g(r,\theta) \tag{4.2}$$

where $r$, $\theta$ are polar coordinates of a point, and $r_1$, $\theta_1$ those of the image point under the mapping $\phi$. Moreover, $f$, $g$ are assumed to be continuous and of period $2\pi$ in $\theta$, and preserving the area

$$\frac{1}{2}\oint r^2\, d\theta \quad . \tag{4.3}$$

THEOREM 2.10. *Let $\phi$ be an area-preserving mapping taking the annulus* (4.1) *homeomorphically into itself mapping the boundaries* $r = a$, $r = b$ *in opposite directions, i.e.,*

$$g(a, \theta)\, g(b, \theta) < 0 \ .^{*}$$

*Then $\phi$ possesses at least one fixed point in the interior of the annulus.*

It is to be noted that this assertion is not true without the area-preserving property and therefore is not a purely topological statement. There are other such fixed point theorems: For example, a continuous, area-preserving mapping of an open disc into itself possesses an interior fixed point; this is clearly incorrect without the area-preserving property.

Applying Theorem 2.10 to the iterates $\phi^p (p = 1, 2, \ldots)$ of $\phi$ one finds at once infinitely many periodic points, which are fixed points of the iterates of $\phi$. Historically this theorem was of interest since such a mapping of an annulus had been constructed for the restricted three-body problem, and thus provided at once infinitely many periodic solutions for this problem.

b) *Invariant curves*

For stability studies it is more important to find invariant curves separating the boundary circles. The existence of such curves can, so far, only be established for small perturbations of so-called twist mappings

---

* Actually $\phi$ defines $g$ in (4.2) only up to an additive integer multiple of $2\pi$ and this condition is meant with some such choice.

(4.4)
$$\begin{cases} r_1 = r \\ \theta_1 = \theta + \alpha(r) \end{cases}$$

which evidently preserve the area (4.3). Furthermore, one has to impose more severe differentiability conditions. Such a twist mapping rotates each circle $r = c$ about an angle $\alpha(c)$, which generally depends on c.

To formulate our existence theorem we introduce for a function $h \in C^{\ell}(A)$ the norm

$$|h|_{\ell} = \sup_{\substack{m+n\leq \ell \\ A}} \left|\frac{\partial^{m+n} h}{\partial r^m \partial \theta^n}\right| .$$

THEOREM 2.11. *Let* $\alpha(r) \in C^{\ell}$ *and* $|\alpha'(r)| \geq \nu > 0$ *in* $a \leq r \leq b$, *for some* $\ell$ *with*

(4.5)
$$\ell \geq 5$$

*and* $\varepsilon$ *be a positive number.*

*Then there exists a* $\delta$ *depending on* $\varepsilon, \ell, \alpha(r)$ *such that any area-preserving mapping* (4.2) *of* A *into* $R^2$* *with* $f, g \in C^{\ell}$ *and*

$$|f-r|_{\ell} + |g-\alpha(r)|_{\ell} < \nu\delta$$

*possesses an invariant curve of the form*

(4.6)
$$r = c + u(\xi) , \quad \theta = \xi + v(\xi)$$

*in* A *where* u, v *are continuously differentiable, of period* $2\pi$ *and satisfy*

$$|u|_1 + |v|_1 < \varepsilon ,$$

*and* c *is a constant in* $a < c < b$. *Moreover, the induced mapping of this curve is given by*

---

* It is not required that this mapping preserves the boundaries.

$$\xi \to \xi + \omega$$

*where* $\omega$ *is incommensurable with* $2\pi$*, and satisfies infinitely many conditions*

$$\left|\frac{\omega}{2\pi} - \frac{p}{q}\right| \geq \gamma q^{-\tau}$$

*with some positive* $\gamma, \tau$*, for all integers* $q > 0$, p. *In fact, each choice of* $\omega$ *in the range of* $\alpha(r)$ *and satisfying the above inequalities gives rise to such an invariant curve.*

One may express this result by saying that the rotation $\xi \to \xi + \omega$ of the circle is embedded via (4.6) as a subsystem of our mapping. Again it is obvious the area-preserving property is essential for this theorem. The above statement ensures that the two annuli into which A is separated by the above curve are invariant sets and hence these mappings are certainly not ergodic. This is in contrast to a celebrated theorem by Oxtoby and Ulam (for references see [64]) which asserts that among the area-preserving homeomorphisms of a disc – measured in a topology related to the maximum norm – the ergodic mappings are of second Baire category, i.e., form the typical case. The above theorem implies that this is certainly not the case in the $C^\ell$-topology if $\ell \geq 5$; the fact that the mapping is defined in an annulus is rather irrelevant in this connection, as we will see below.

The condition (4.5) is certainly not sharp. At first this theorem was proven for $\ell \geq 333$ [63], it was improved by Rüssman to $\ell \geq 5$ and, in fact, using the arguments of [65], [60] it suffices to use $\ell > 3$ where $C^\ell$ for non-integer $\ell = m + \mu$, m integer, $0 \leq \mu < 1$ means that the $m^{th}$ derivatives are Hölder continuous with Hölder exponent $\mu$, and $|h|_\ell$ denotes the corresponding Hölder norm. The resulting curves have still continuous derivatives, in fact, they are Hölder continuous. One may conjecture that for $\ell > 2$ one still obtains the existence of continuous invariant curves, however, this has not been proven. On the other hand, it was shown by F. Takens [67] that for $\ell = 1$ the statement of the above theorem is false, by showing that one can construct a sequence of iterates which come close to $r = a$ as well as $r = b$. Thus $\ell$ has to be chosen greater than 1; this phenomenon is reminiscent and probably related to a theorem of Denjoy [68] on circle maps which require the mapping to have first derivatives of bounded variations but is false for $C^1$-mappings.

Actually the above theorem yields an infinite number of invariant curves, and it can be shown that the set of such invariant curves omits only a set of arbitrarily small measure, if $\delta$ is taken small enough. This has the curious implication that the closure of the periodic points is a set of positive measure, since it can be shown from Theorem 2.10 and Theorem 2.11 that the closure of the periodic points of such a mapping contains the set of these invariant curves.

c) *Elliptic fixed points*

The above theorem can be applied to a great number of problems. We mention a few such applications briefly, and refer to [8] for a complete treatment.

We consider a sufficiently often differentiable area-preserving mapping $\phi$ in the plane and discuss $\phi$ in the neighborhood of a fixed point. These fixed points are classified according to the eigenvalues $\lambda_1, \lambda_2$ of the Jacobian mapping $d\phi$ at this fixed point. Since the determinant of the Jacobian is one it follows that $\lambda_1\lambda_2 = 1$, and since the mapping is real one has $\overline{\lambda}_1 = \lambda_1$, $\overline{\lambda}_2 = \lambda_2$ or $\overline{\lambda}_1 = \lambda_2 = \lambda_1^{-1}$. Thus in the second case $|\lambda_1| = 1$, while in the first $\lambda_1, \lambda_2$ are real. If the eigenvalues are assumed to be distinct one speaks in the first case of two real eigenvalues $\neq \pm 1$ of a hyperbolic fixed point, and in the second, when $|\lambda_1| = 1$, $\lambda_2 = \overline{\lambda}_1 \neq \pm 1$ of an elliptic fixed point. Finally, the exceptional case of two equal eigenvalues $\pm 1$ is called parabolic.

The nature of real analytic mapping near such fixed points was discussed in the long paper of G. D. Birkhoff [62] who constructed normal forms for such mapping analogue to those of Section 2. However, these results are all formal and give rise, in general, to divergent series. Here we mention a stability result for such fixed points which was not accessible to the power series approach.

A fixed point can be stable only in the elliptic case, when the eigenvalues $\lambda$, $\bar{\lambda}$ lie on the unit circle, if we omit the parabolic case. After a linear transformation one may assume that the fixed point is at the origin and the mapping is of the form

$$z \to z_1 = f(z, \bar{z}) = \lambda z + g(z, \bar{z}) \tag{4.7}$$

where $z = x + iy$, $\bar{z} = x - iy$ are complex variables and $g$ vanishes with its first derivatives at $z = 0$. Birkhoff's normal form for this situation is described by the following theorem:

THEOREM 2.12. *If* (4.7) *is an area-preserving mapping in* $C^\ell$ $(\ell \geq 4)$ *and if with some integer* $q$ *in*

$$4 \leq q \leq \ell+1$$

*one has*

$$\lambda^k \neq 1 \quad \textit{for} \quad k = 1, 2, \ldots, q$$

*then there exists a real analytic canonical transformation* $(x, y) \to (\xi, \eta)$ *taking* (4.6) *into*

$$\zeta \to \zeta_1 = \lambda \zeta e^{ia(\zeta\bar{\zeta})} + h(\zeta, \bar{\zeta}) \tag{4.8}$$

*where* $\zeta = \xi + i\eta$,

$$a(\zeta\bar{\zeta}) = \alpha_1 |\zeta|^2 + \ldots + \alpha_s |\zeta|^{2s}, \quad s = \left[\frac{q}{2}\right] - 1 \tag{4.9}$$

*is a real polynomial in* $|\zeta|^2$ *and* $h$ *vanishes with its derivatives up to order* $q-1$ *at* $\zeta = \bar{\zeta} = 0$.

Formula (4.8) shows that near an elliptic fixed point our mapping can be approximated by

$$\zeta_1 = \lambda \zeta e^{ia(\zeta\bar{\zeta})}$$

which is, in fact, a twist mapping with $a(\zeta\bar{\zeta})$ playing the role of $\alpha(r)$ in (4.4). Therefore if $a(\zeta\bar{\zeta})$ is not identically zero one has $\alpha'(r) \neq 0$ and Theorem 2.11 is applicable and gives

THEOREM 2.13. *If for an area-preserving mapping of the form* (4.8) *the polynomial* $a(|\zeta|^2)$ *in* (4.9) *does not vanish identically then* $\zeta = 0$ *is a stable fixed point.*

The proof is obtained by applying Theorem 2.11 to an annulus $\varepsilon < |\zeta| < 2\varepsilon$, where $\varepsilon$ is a small positive number. This yields an invariant curve surrounding the origin, and its interior is an invariant set containing the origin. This way one finds such an open invariant set in any neighborhood of $\zeta = 0$, guaranteeing stability.

The significance of this result is that the stability criterion is expressed in terms of finitely many inequalities. In particular, one has to exclude only a finite number of roots of unity to ensure stability. The typical case occurs when one takes $q = 4$, $s = 1$ and $a_1 \neq 0$ in Theorem 2.12, so that $\lambda$ has to avoid the 6 roots of unity of order 1, 2, 3, 4.

If the above conditions are violated one can construct counterexamples. Particularly interesting is a recent result of Anosov and Katok [69]. They constructed examples of area-preserving mappings of class $C^\ell$, $\ell$ arbitrary, taking the disc $|z| < 1$ into itself and having $z = 0$ as an elliptic fixed point, but which are ergodic in $|z| < 1$. Thus these mappings have no open invariant set except for $|z| < 1$ and hence $z = 0$ is certainly not a stable fixed point. In their examples, however, the polynomial (4.9) vanishes identically, showing the necessity of such a condition. At the same time it is clear that a small perturbation will destroy the ergodicity of such a mapping, and these examples are in this sense exceptional.

d) *Applications to the plane restricted three-body problem*

Theorem 2.12 has many applications, in particular for the study of stability of periodic solutions of Hamiltonian systems of two degrees of freedom. This is clear if one recalls (see a)) that one can associate with the Hamiltonian flow near a periodic orbit an area-preserving mapping. Verification of the finitely many conditions for this mapping would ensure stability of the periodic orbits.

In this manner one can completely decide about the stability of the so-called periodic orbits of the first kind of the restricted 3-body problem. This is formulated as follows: One considers 2 mass points $P_1$, $P_2$ of masses $m_1, m_2 \geq 0$ moving on circles about their common center of mass. In the plane of motion of these so-called primaries one considers a third mass point $P_3$ of mass $m_3 = 0$ which therefore does not affect the primaries and the problem is to study the motion of the zero mass point $P_3$. Usually one studies only the case of small mass ratio $\mu = \frac{m_2}{m_1 + m_2}$.

For $\mu = 0$ clearly the problem reduces to the two-body problem as the second mass point drops out of consideration and we consider the circular orbits for $P_3$ about $P_1$. They form a one-parameter family, which may be parametrized by the frequency $\nu_3$ of rotation of $P_3$ about $P_1$.

If one introduces a coordinate system, rotating with the frequency $\nu_2$ of $P_2$, in which $P_1$, $P_2$ are at rest the circular orbit of $P_3$ has the frequency $\nu_3 - \nu_2$, and the period $(\nu_3 - \nu_2)^{-1} 2\pi$. The periodic solutions of the first kind are solutions of the restricted three-body solution for small but positive values of $\mu$ which for $\mu \to 0$ tend to the above circular orbits, and it is well known [8] that such orbits, exist for sufficiently small $\mu$, if $\nu_3 \neq \pm \nu_2$ and

$$\frac{\nu_2}{\nu_3 - \nu_2} \neq m$$

for all integers m.

To decide about the stability about these orbits one constructs an area-preserving $\phi = \phi(\mu)$ for this flow and investigates the conditions of Theorem 2.12. In fact, as all quantities $\lambda$, $a_1, \ldots, a_s$ involved depend continuously on $\mu$ it suffices to verify these conditions for $\mu = 0$, in which case all quantities are computable. It turns out that for $\mu = 0$ the conditions of Theorem 2.12 hold with $q = 4$, $\ell = \infty$ provided

$$\frac{\nu_2}{\nu_3 - \nu_2} \neq \frac{m}{n} \quad \text{for} \quad n = 1, 2, 3, 4 \tag{4.10}$$

for all integers m. Under this condition the periodic orbits of the first kind are stable, if $\mu$ is sufficiently small.

Actually the above conditions are only sufficient conditions and it turns out that the above condition for $n = 4$ can be dropped. However, for

$$\frac{\nu_2}{\nu_3 - \nu_2} = \frac{1}{3}$$

one can verify that one has instability for small $\mu > 0$.

The resulting stability conditions (4.10) for $n = 1, 2, 3$ reflect an astronomical phenomenon, the so-called Kirkwood gaps of the asteroids, which are most pronounced when these stability conditions are violated (see [8], [70]).

Theorem 2.11 can also be used to establish a finite set of inequalities ensuring stability of an equilibrium for Hamiltonian systems of two degrees of freedom. In Theorem 2.8 we mentioned such conditions for systems of arbitrary degrees of freedom which, however, yielded only a weaker type of stability. However, for 2 degrees of freedom it is possible to give conditions for actual stability in the sense of Liapunov, without exceptional orbits. Such conditions can be applied to assert the stability of the triangular Lagrange solution of the restricted three-body problem in which the three mass points move in a rotating equilateral triangle. For more than two degrees of freedom it is still an unsolved problem whether one can find a finite set of inequalities imposed on a finite number of Taylor coefficients of the Hamiltonian which ensure stability of the equilibrium except if $\pm H$ has a local minimum. If this is not possible one would have to consider Theorem 2.8 as an acceptable substitute.

We mention that Theorem 2.11 allows us to show that for all solutions of Ulam's example mentioned in Chapter I the velocity remains bounded for all time. Similarly, for the geometrical problem, based on the construction of successive tangents to a convex curve, all orbits can be shown to be bounded, if the curve has at least 6 continuous derivatives.

There are many other applications of the above theorems, or their generalizations, to various problems of mechanics and other areas of physics [71-76]. The results of this chapter provide quasi-periodic solutions with rationally independent frequencies, and their existence can be established in spite of the presence of the small divisors. In order to give the basic idea of how to overcome the difficulty of the small divisors we provide the proof of Theorem 2.9 in Chapter V.

# CHAPTER III

# STATISTICAL BEHAVIOR

## 1. *The Bernoulli shift. Examples*

a) In contradistinction to the previous chapter we will now study orbits which are quite unpredictable in their behavior for long time intervals. We begin with some well-known and simple examples from ergodic theory or probability theory, the so-called Bernoulli shifts, which are defined over a measure space, discuss then the corresponding topological transformation, the shift of doubly infinite sequences and finally embed it into the three-body problem. In actual fact this example is not an isolated one but this phenomenon occurs quite generally for Hamiltonian system of differential equations as we will see at the end of this chapter. However, the geometrical discussion is particularly transparent and interesting for the example of the three-body problem that we chose for illustration. The full details and computations are supplied in Chapter VI.

The idea to characterize orbits by a sequence of symbols or integers has a long history. It occurs already in the study of Hadamard [77], Birkhoff [4], [89] and of Morse and Hedlund [78] on geodesics on some surfaces on negative curvature. In a completely different context such an approach occurred in the work of Cartwright and Littlewood as well as Levinson [79] on some periodically excited van der Pol equations which occur in electrical circuits. They found situations where one could prescribe at infinitely many stages through which of two "gates" the solutions should pass, thus effectively describing these solutions by an infinite sequence of zeros and ones. In recent times these phenomena were cast into a very geometrical and useful form by Smale [80]. A systematic study of these phenomena is due to Alekseev [34],[35] who made a beautiful application of these ideas to the restricted three-body problem completing a previous work of Sitnikov [33] on this model. At the same time Conley [82-84] obtained similar results establishing, for example, motions for the restricted three-body problems which oscillate infinitely often between the two primaries.

In the following we give an exposition of these ideas applying them to the example of Sitnikov. The calculations and details in the cited paper of Alekseev are formidable and difficult to follow and we hope that the exposition given here simplifies the matter. The main ideas of this approach were developed at a joint seminar held by Conley and the author at New York University in 1969-1970 and the exposition is largely that of Conley. In particular, a clever device suggested by McGehee will be decisive in simplifying the geometrical constructions and the calculations.

b) Let $A$ denote a finite or denumerable set which we will call an alphabet. The doubly infinite sequence

$$s = (\dots s_{-2}, s_{-1}, s_0; s_1, s_2, \dots)$$

of elements $s_k \epsilon A$ form the elements of a space. We introduce a topology in $S$ by taking as neighborhood basis of

$$s^* = (\dots, s^*_{-1}, s^*_0; s^*_1, s^*_2, \dots)$$

the sets

$$U_j = \{s \epsilon S \mid s_k = s^*_k(|k| < j)\}$$

for $j = 1, 2, \dots$ . This makes $S$ a topological space. The shift homeomorphism $\sigma$ is defined on $S$ by

$$(\sigma(s))_k = s_{k-1} \ .$$

This mapping will play a basic role in the following.

The same type of mapping is often considered from a measure theoretical point of view. For this purpose we make $S$ into a measure space: Assigning to every element $a \epsilon A$ a positive number $p_a$ as its measure, so that

$$\sum_{a \in A} p_a = 1$$

we define the product measure in the standard fashion, by defining the measure of the sets

$$E_{k_1 \ldots k_r}(a_1, \ldots, a_r) = \{s \in S;\ s_{k_1} = a_1, \ldots, s_{k_r} = a_r\}$$

by $p_{a_1} p_{a_2} \cdots p_{a_r}$ and extending it to the Borel algebra $\Sigma$ generated by the above sets. Then $\{S, \Sigma, m\}$ with the so defined measure $m$ forms a measure space and the shift which we now denote by $\sigma_m$ obviously preserves this measure. The mapping $\sigma_m$ is the so-called Bernoulli shift. It is well known that $\sigma_m$ is ergodic – i.e., every Borel set which is invariant under $\sigma_m$ is of measure zero or one – and even mixing.

In the special case where $A = \{0, 1\}$ and $p_0 = p_1 = \frac{1}{2}$ we give $\sigma_m$ a more geometric interpretation: Associating with any $s \in S$ two binary numbers

$$x = \sum_{k=0}^{-\infty} s_k 2^{k-1}, \quad y = \sum_{k=1}^{\infty} s_k 2^{-k}$$

we obtain a mapping $\tau$ of $S$ into the square

$$Q: \quad 0 \le x \le 1, \quad 0 \le y \le 1 .$$

Moreover, $\tau$ takes the measure $m$ into the Lebesgue measure in $Q$ so that $\phi = \tau\sigma\tau^{-1}$ is a measure preserving transformation in $Q$. This mapping is easily identified with $(x, y) \to (x_1, y_1)$ where

$$x_1 = 2x - [2x], \quad y_1 = \frac{1}{2}(y + [2x])$$

with $[x]$ denoting the largest integer $\le x$. This is sometimes called the "baker transformation" since it resembles the process of rolling out the dough: The two strips $V_1 = \{0 \le x < \frac{1}{2}\}$, $V_2 = \{\frac{1}{2} \le x \le 1\}$ are mapped by $\phi$ into $U_1 = \{0 \le y < \frac{1}{2}\}$ and $U_2 = \{\frac{1}{2} \le y \le 1\}$, respectively. Thus the rectangles get stretched by a factor 2 in the horizontal direction and compressed by a factor $\frac{1}{2}$ in the vertical direction (Fig. 3).

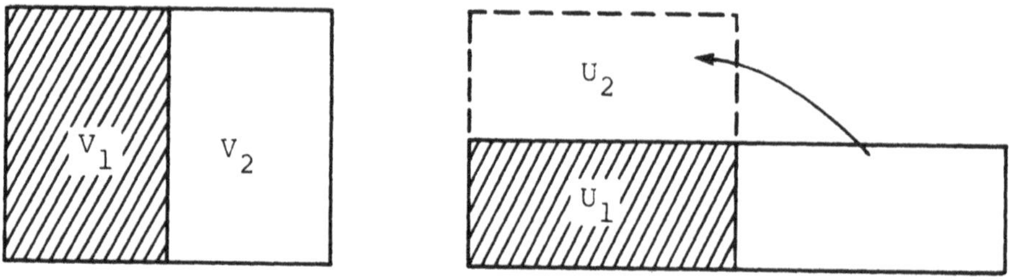


Fig. 3

Alternately $V_1$, $V_2$ can be described as the set of $(x, y)$ with $s_0 = 0, 1$ respectively. They are mapped into the sets where $s_1 = 0, 1$, respectively.

Since the baker transformation $\phi$ is isomorph with $\sigma$, via $\tau$, it is also ergodic and even mixing.

Similarly, as the above example is based on subdivision of the square $Q$ into two rectangles $V_1$, $V_2$ one can obviously construct similar examples by cutting into finitely many, say $N$ vertical rectangles and relate the mapping to Bernoulli shift of $N$ elements.

A more interesting mapping of the interior of $Q$ is given by

$$x_1 = \frac{1}{x} - \left[\frac{1}{x}\right], \quad y_1 = \left(y + \left[\frac{1}{x}\right]\right)^{-1}$$

which preserves the measure

$$\frac{1}{\log 2} \frac{dxdy}{(1+xy)^2} .$$

It is related to a shift of doubly infinite sequence of positive integers, i.e., $N = \infty$. Indeed, if one represents $x$, $y$ in terms of continued fractions, via

$$x = \cfrac{1}{s_0 + \cfrac{1}{s_{-1} + \cfrac{1}{s_{-2} + \dots}}} ; \quad y = \cfrac{1}{s_1 + \cfrac{1}{s_2 + \dots}}$$

where $s_k$ are positive integers, the image point $x_1$, $y_1$ is given by the shifted sequence $\sigma(s)$ where $(\sigma(s))_k = s_{k-1}$. Actually this is *not* a Bernoulli shift because the transformed measure is not a product measure. Nevertheless it is true that also this mapping is ergodic and mixing. It is even possible to compute the entropy of this mapping to be $\frac{\pi^2}{6 \log 2}$ [29].

This example plays a role in the study of the geodesic flow on the modular region which was studied in an interesting paper by Artin [85] in 1924.

c) The above examples are given by non-continuous mappings and we mention as third example a continuous mapping $\phi$ of the torus onto itself. We describe the torus by identifying the points x, y in the plane whose coordinates differ by integers. If A is a matrix with integer coefficients the mapping $z \to Az$ takes equivalent points into equivalent ones, i.e., the above defines a mapping $\phi$ on the torus. If $\det A = \pm 1$ the same is true for the inverse mapping and $\phi$ preserves the Lebesgue measure. If the eigenvalues of A are not on the unit circle this mapping is mixing. In fact, recently I. Katznelson showed that this mapping $\phi$ is even isomorph with the Bernoulli shift.

For an ergodic mapping, continuous with respect to a topology for which open sets have positive measure for almost all p in the measure space the orbits $\{\phi^k(p)\}$ $(k = 0, \pm+, \pm 2, \ldots)$ are dense. For Bernoulli shifts also the set of periodic points, i.e., points p for which $\phi^m(p) = p$ for some $m \geq 1$, are dense. This latter property is readily verified for our example. The set of points with rational coordinates with a fixed denominator, say q, is mapped into itself by the above example. Since there are only $q^2$ such points on the torus these must be periodic. Thus all points with rational coordinates are periodic points, and they are clearly dense. Similarly, for a Bernoulli shift the periodic points are given by periodic sequences; they are clearly dense in S as one sees by replacing one arbitrary sequence by repeating a sufficiently large section of the sequence.

2. *The shift as a topological mapping*

a) Usually the shift is considered from the *measure theoretical* point of view, as in the above examples. In the following we will consider the shift merely as a *topological* mapping, since so far one has only few examples of differential equations where one can relate the orbits to the shift on a measure space. On the other hand the shift as a topological mapping occurs quite frequently for nonlinear differential equations, as we intend to show.

b) First, following Smale [80, 81], we shall relate the shift, now considered as a topological mapping, to a geometrical mapping on a square Q, similarly, as we did in the previous examples. To do this informally, we assume that we have a topological mapping $\phi$ of the closed square Q into the plane, such that $\phi(Q)$ intersects Q in two components, denoted by $U_1$, $U_2$ (see Fig. 4), with the indicated boundary correspondence.

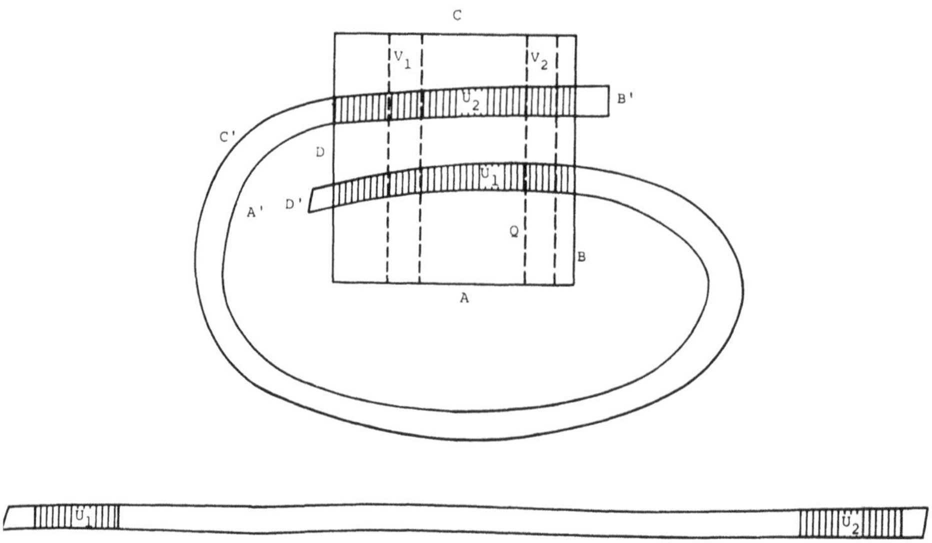


Fig. 4

Then the preimages of $U_1$, $U_2$ will be denoted by $V_a = \phi^{-1}(U_a)$, $a = 1, 2$. They correspond to vertical strips connecting the upper and lower edges of Q, and

$$V_1 \cup V_2 = Q \cap \phi^{-1}(Q) .$$

Obviously the iterates $\phi^k$ of $\phi$ are not defined in all of Q and we construct first the invariant set

$$I = \bigcap_{k=-\infty}^{+\infty} \phi^{-k}(Q) \tag{2.1}$$

in which all iterates $\phi^k$ are, in fact, defined. Of course, it remains to be shown that I is not empty. Postponing the detailed proofs of this and the following statements to the next section we associate with a point $p \in I$ a doubly infinite sequence of symbols in $A = \{1, 2\}$ by the rule

$$\phi^{-k}(p) \in V_{s_k} \qquad (k = 0, \pm 1, \pm 2, \ldots) \tag{2.2}$$

or

$$p \in \bigcap_{k=-\infty}^{+\infty} \phi^k(V_{s_k}) . \tag{2.3}$$

This provides a mapping of I into the sequence space S. Under appropriate conditions we shall show that this mapping has an inverse $\tau : S \to I$, i.e., to every sequence $s \in S$ one can find a $p \in I$ with the above property, and that $\tau$ is a homeomorphism. From the definition of $\tau$ it is evident that then one has

$$\phi^{-k}(\phi^{-1}(p)) \in V_{s_{k+1}}$$

i.e., $\tau^{-1}\phi|_I\tau = \sigma$, showing that $\phi|_I$ is topologically equivalent to $\sigma$. In the following section we will make these statements precise.

If $\phi$ is a topological mapping of the topological space T into itself and $\psi$ a second such mapping of the space S into itself, we call the latter a subsystem of the form if there exists a homeomorphism $\tau$ of S into $\tau(S) \subset T$ such that

(2.4) $$\phi\tau = \tau\psi \ .$$

Our aim is then to show that the shift $\sigma$ is a subsystem of $\phi$. This is to be understood as a reflection of the statistical behavior of $\phi$: If one knows into which strips $V_{s_k}$ the $\phi^k(p)$ fall for $|k| < K$ one cannot make any prediction about the later behavior of the orbit, no matter how large $K$ is. In fact, the numbers $s_k$ can be chosen quite independently in $A = \{1, 2\}$.

Of course, one can deduce other facts about $\phi$. For example, the existence of infinitely many periodic points follows immediately since $\sigma$ is a subsystem of $\phi$, and, moreover, that the periodic points are dense in $I$.

3. *The shift as a subsystem*

a) To formulate the above statement more precisely we introduce some concepts: Given a number $\mu$ in $0 < \mu < 1$ we call a curve $y = u(x)$ a horizontal curve if $0 \leq u(x) \leq 1$ for $0 \leq x \leq 1$ and

(3.1) $$|u(x_1) - u(x_2)| \leq \mu |x_1 - x_2| \quad \text{in} \quad 0 \leq x_1 \leq x_2 \leq 1 \ .$$

If $u_1(x)$, $u_2(x)$ define two such horizontal curves and if

$$0 \leq u_1(x) < u_2(x) \leq 1$$

we call the set

$$U = \{(x,y) \mid 0 \leq x \leq 1; \ u_1(x) \leq y \leq u_2(x)\}$$

a horizontal strip and denote by $\mathfrak{U} = \mathfrak{U}(\mu)$ the collection of all these horizontal strips. Finally,

(3.2) $$d(U) = \max_{0 \leq x \leq 1} (u_2(x) - u_1(x))$$

will be called the diameter of $U$ (see Fig. 5).

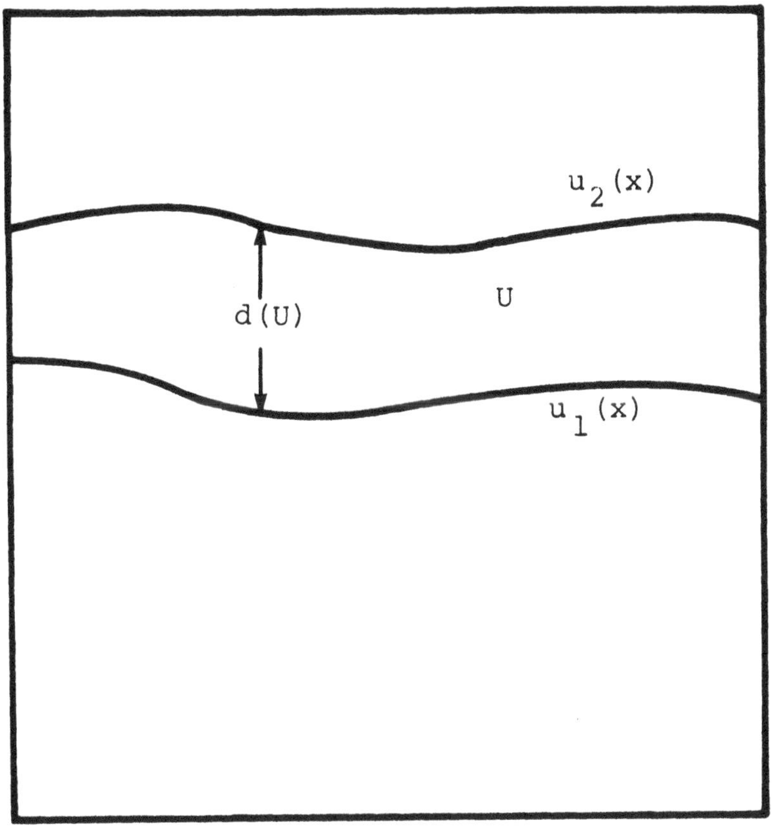


Fig. 5

Similarly, we describe a vertical curve by $x = v(y)$ where for $0 \leq y \leq 1$

$$(3.3) \quad 0 \leq v(y) \leq 1, \ |v(y_1) - v(y_2)| \leq \mu |y_1 - y_2| \quad \text{in} \quad 0 \leq y_1 \leq y_2 \leq 1 .$$

A vertical strip $V$ is given by two vertical curves and the above definitions carry over by exchanging the role of $x$ and $y$. The set of all vertical strips will be denoted $\mathfrak{V} = \mathfrak{V}(\mu)$; the number $\mu (0 < \mu < 1)$ will be fixed throughout.

b) In the following we shall need some elementary facts about these strips:

LEMMA 1. *If* $U^{(1)} \supset U^{(2)} \supset U^{(3)} \supset \ldots$ *is a sequence of nested horizontal strips* $U^{(k)} \in \mathfrak{U}(k = 1, 2, \ldots)$ *and if* $d(U^{(k)}) \to 0$ *as* $k \to \infty$ *then*

$$\bigcap_{k=1}^{\infty} U^{(k)}$$

*defines a horizontal curve.*

The proof of this lemma is readily established using the compactness property of the family of horizontal curves. A similar statement holds, of course, for nested sequences in $\mathfrak{B}$.

LEMMA 2. *A horizontal curve* $y = u(x)$ *and a vertical one,* $x = v(y)$ *intersect in precisely one point.*

*Proof.* A point of intersection $(x, y)$ is characterized by a zero $x$ of $x - v(u(x))$ and $y = u(x)$. Since by (3.1), (3.3), for $0 \leq x_1 < x_2 \leq 1$, one has

$$|v(u(x_1)) - v(u(x_2))| \leq \mu |u(x_1) - u(x_2)| \leq \mu^2 |x_1 - x_2|$$

and $\mu^2 < 1$ the function $x - v(u(x))$ is strictly monotonically increasing. Since it is $\leq 0$ for $x = 0$ and $\geq 0$ for $x = 1$ it has precisely one zero which proves the statement.

According to Lemma 2 we have a mapping of the space of pairs of curves (one horizontal and one vertical) into $Q$ by assigning to the pair $(u, v)$ the point of intersection, say $z = (x, y)$. This mapping is Lipschitz continuous in the norms

$$\|u\| + \|v\| = \max_{0 \leq x \leq 1} |u(x)| + \max_{0 \leq y \leq 1} |v(y)|$$

and

$$|z| = |x| + |y| .$$

In fact, if $z_j$, $(j = 1,2)$ correspond to $(u_j, v_j)$, $j = 1,2$ one has

$$|z_2 - z_1| \leq (1-\mu)^{-1}(\|u_2 - u_1\| + \|v_2 - v_1\|) \ . \tag{3.4}$$

Indeed, since $x_j = v_j(y_j)$ one has by (3.3)

$$|x_2 - x_1| \leq |v_2(y_2) - (v_1(y_2)| + |v_1(y_2) - v_1(y_1)| \leq \|v_2 - v_1\| + \mu|y_2 - y_1| \ .$$

Similarly, one finds from (3.1)

$$|y_2 - y_1| \leq \|u_2 - u_1\| + \mu|x_2 - x_1|$$

and addition of these inequalities yields the statement (3.4), since $0 < \mu < 1$.

c) We come to the formulation of the assumptions for the mapping $\phi$ defined in Q:

i) Let A be the set $\{1, 2, \ldots, N\}$ if $N < \infty$ or the set of positive integers if $N = \infty$ and assume that $U_a$, $V_a$ for $a \in A$ are given as disjoint horizontal, respectively, vertical strips in Q. The mapping $\phi$ takes $V_a$ homeomorphically into $U_a$, i.e.,

$$\phi(V_a) = U_a \ , \quad a \in A \ . \tag{3.5}$$

Moreover, it is required that the vertical boundaries of $V_a$ are mapped onto the vertical boundaries of $U_a$; similarly the horizontal boundaries should correspond under the mapping $\phi$.

ii) If V is a vertical strip in $\bigcup_{a \in A} V_a$ then, for any $a \in A$

$$\phi^{-1}(V) \cap V_a = \tilde{V}_a \tag{3.6}$$

is a vertical strip (in particular non-empty) and for some fixed $\nu$ in $0 < \nu < 1$ we require

$$d(\tilde{V}_a) \leq \nu \, d(V_a) \ . \tag{3.7}$$

Similarly, if U is a horizontal strip in $\bigcup_{a \in A} U_a$

(3.8) $$\phi(U) \cap U_a = \tilde{U}_a \; ^{*}$$

is assumed to be a horizontal strip with

(3.9) $$d(\tilde{U}_a) \leq \nu \, d(U_a) \; .$$

THEOREM 3.1. *If* $\phi$ *is a homeomorphism satisfying the above hypotheses* i) *and* ii) *with respect to the horizontal and vertical strips* $U_a$, $V_a (a \in A)$ *then it possesses the shift* $\sigma$ *on the sequences of elements of* A *as a subsystem. That is, there exists a homeomorphism* $\tau$ *of* S *into* Q *such that*

$$\phi\tau = \tau\sigma \; .$$

In particular, if $N < \infty$, $\tau(S)$ is a closed and invariant set in Q which, like S, is a Cantor set.

d) *Proof*: In order to study the set of points $p \in Q$ for which $\phi^{-k}(p) \in V_{s_k} (k = 0, \pm 1, \pm 2, \ldots)$ we define inductively for $n \geq 1$

(3.10) $$V_{s_0 s_{-1} \cdots s_{-n}} = V_{s_0} \cap \phi^{-1}(V_{s_{-1} \cdots s_{-n}})$$

and verify inductively, using assumption ii), that these are vertical strips. Also, by the same assumption, the diameter

$$d(V_{s_0 s_{-1} \cdots s_{-n}}) \leq \nu \, d(V_{s_{-1} \cdots s_{-n}}) \leq \nu^n d(V_{s_{-n}}) \leq \nu^n$$

of $V_{s_0 s_{-1} \cdots s_{-n}}$ tends to zero as $n \to \infty$. From the definition of $V_{s_0 s_{-1} \cdots s_{-n}}$ one sees that they agree with the sets

* By $\phi(U)$ is meant the image of $U \cap \bigcup_{a \in A} V_a$.

(3.11) $$V_{s_0 s_{-1} \dots s_{-n}} = \{p \in Q, \phi^k(p) \in V_{s_{-k}} \ (k = 0, 1, \dots, n)\}$$
$$= \{p \in Q, \phi^{-k}(p) \in V_{s_k} \ (k = 0, -1, \dots, -n)\} ,$$

and hence

$$V_{s_0 s_{-1} \dots s_{-n}} \subset V_{s_0 s_{-1} \dots s_{-n+1}} .$$

Thus, by Lemma 1, the intersection

(3.12) $$V(s) = \bigcap_{n=0}^{\infty} V_{s_0 s_{-1} \dots s_{-n}} = \{p \in Q, \phi^{-k}(p) \in V_{s_k} \ (k = 0, -1, \dots)\}$$

defines a vertical curve, depending on the left half of the sequence $s$.

Similarly, we define inductively for $n \geq 2$

(3.13) $$U_{s_1 s_2 \dots s_n} = U_{s_1} \cap \phi(U_{s_2 \dots s_n})$$

which by assumption ii) are nested horizontal strips whose diameters can be estimated by $\nu^{n-1}$. Thus, since $\phi(V_{s_k}) = U_{s_k}$

(3.14) $$U(s) = \bigcap_{n=1}^{\infty} U_{s_1 s_2 \dots s_n} = \{p \in Q, \phi^{-k+1}(p) \in U_{s_k}, \ k \geq 1\}$$
$$= \{p \in Q, \phi^{-k}(p) \in V_{s_k}, \ k \geq 1\}$$

is a horizontal curve, depending on the right half of $s$. By Lemma 2 the intersection

(3.15) $$V(s) \cap U(s) = \{p \in Q \mid \phi^{-k} \in V_{s_k} \ (k = 0, \pm 1, \pm 2, \dots)\}$$

defines precisely one point in $Q$. Now we define the mapping $\tau$ of $S$ into $Q$ by assigning to the sequence $s = (\dots s_{-1} s_0 s_1, \dots) \in S$ the above point $V(s) \cap U(s)$.

From the construction it is clear that if $\tau(s) = p$ then the shift sequence $\sigma(s)$ is mapped into $\tau\sigma(s) = \phi(p)$, hence

$$\tau\sigma = \phi\tau .$$

It remains to be shown that $\tau$ is continuous, that it is injective and its inverse also continuous.

The continuity of $\tau$ follows immediately from the construction: If s, s′ agree in its $k^{th}$ components for $|k| \leq n$, then the image points $\tau(s)$, $\tau(s')$ lie in the same vertical strip $V_{s_0 s_1 \dots s_n}$ and the same horizontal strip $U_{s_1 \dots s_n}$. Since $d(V_{s_0 s_1 \dots s_n}) \leq \nu^n$, $d(U_{s_1 \dots s_n}) \leq \nu^{n-1}$ implies

$$|\tau(s) - \tau(s')| \leq (1-\mu)^{-1}(\nu^n + \nu^{n-1})$$

which shows the continuity of $\tau$.

The fact that the strips $V_a$ are disjoint implies immediately that $\tau$ is injective. Obviously, the image of $\tau$ is precisely the set

$$\bigcap_{k=-\infty}^{+\infty} \phi^k(W) \quad \text{if} \quad W = \bigcup_{a \in A} V_a ,$$

and therefore a compact subset of Q if $N < \infty$.

Since $\tau$ is injective we conclude that $\tau^{-1}$ is continuous, if $N < \infty$. Actually, the same is true for $N = \infty$ but we omit the proof. This completes the proof of Theorem 3.1.

e) In the case $N = \infty$ the set S, and hence $\tau(S)$ is not compact. We discuss a compactification of S which we will use later on. We assume $A = \{1, 2, \dots\}$ and assume that $V_a$ is ordered according to increasing abscissas with increasing a. Similarly we assume that $U_{a+1}$ lies above $U_a$. Then it is clear that $V_a$ tends to a vertical curve $V_\infty$ which we may assume agrees with $x = 1$; similarly we assume that the limit $U_\infty$ of the horizontal curves is $y = 1$. The difficulty arises from the fact that $\phi$ is, in general, not defined on $V_\infty$, and $\phi^{-1}$ not on $U_\infty$. For this reason the compactification of $\tau(S)$ will not any more be an invariant set of $\phi$.

To introduce the compactification $\overline{S}$ of S we admit elements of the following type: For $\kappa$, $\lambda$ are integers satisfying $\kappa \leq 0$, $\lambda \geq 1$ let

$$s = (\infty, s_{\kappa+1}, \ldots, s_{\lambda-1}, \infty); \; s_k \in A \tag{3.16}$$

where $\kappa = 0$, $\lambda = 1$ corresponds to the symbolic element $(\infty, \infty)$. If $\kappa = -\infty$, $\lambda = +\infty$ we identify the above sequences with the elements in S. We also admit half infinite sequences with $\kappa = -\infty$, $\lambda < \infty$ or $\kappa > -\infty$, $\lambda = \infty$. These sequences form the elements of the space $\overline{S} \supset S$. We introduce as neighborhoods of the element

$$s^* = (\ldots s^*_{-1}, s^*_0, \ldots, s^*_{\lambda-1}, \infty)$$

with $\kappa = -\infty$, $\lambda < \infty$ the set of s with

$$s_k = s^*_k \quad \text{for} \quad -K \le k < \lambda$$

$$s_\lambda \ge K$$

for $K = 1, 2, 3, \ldots$ . Defining for the other types the neighborhoods analogously we obtain a topology in $\overline{S}$, extending that of S, in which $\overline{S}$ is compact.

The shift $\sigma$ is not any more defined in all of S. We extend $\sigma$ to $\overline{\sigma}$ which is the shift formally defined as before, in the domain

$$D(\overline{\sigma}) = \{s \in \overline{S}, \; s_0 \neq \infty\} .$$

The range $R(\overline{\sigma})$ of $\overline{\sigma}$ is then given by

$$R(\overline{\sigma}) = \{s \in \overline{S}, \; s_1 \neq \infty\} .$$

It is not difficult to extend the homeomorphism $\tau$ of Theorem 3.1 to a homeomorphism $\overline{\tau}$ of $\overline{S}$ into Q such that

$$\overline{\tau}\,\overline{\sigma} = \phi\overline{\tau}|_{D(\overline{\sigma})} .$$

The set $\overline{\tau}(\overline{S})$ is compact, but not anymore invariant under $\phi$. The new states (3.16) correspond to solutions which "escape" for positive or negative time.

4. *Alternate conditions for* $C^1$*-mappings*

a) The above conditions, in particular ii), are still hard to verify and we will replace ii) by a different condition, in the case $\phi$ is continuously differentiable. Representing $\phi$ in coordinates by

$$x_1 = f(x_0, y_0) \tag{4.1}$$

$$y_1 = g(x_0, y_0)$$

where $(x_1, y_1)$ is the image point of $(x_0, y_0)$ we associate with it the mapping $d\phi$ taking the tangent vector $(\xi_0, \eta_0)$ at $(x, y)$ into the vectors $(\xi_1, \eta_1)$ at $(x_1, y_1)$ where

$$\begin{cases} \xi_1 = f_x \xi_0 + f_y \eta_0 \\ \eta_1 = g_x \xi_0 + g_y \eta_0 \ . \end{cases} \tag{4.2}$$

Now we replace ii) in Section 3 by the following assumption:

iii) For some $\mu$ in $0 < \mu < 1$ the bundle of sectors

$$S^+ : |\eta| \le \mu |\xi| \tag{4.3}$$

defined over $\bigcup_{a \epsilon A} V_a$, is mapped into itself by $d\phi$, i.e.,

$$d\phi(S^+) \subset S^+ \ . \tag{4.4}$$

Moreover, if $(\xi_0, \eta_0) \ \epsilon \ S^+$ and $(\xi_1, \eta_1)$ its image point, then

$$|\xi_1| \ge \mu^{-1} |\xi_0| \ . \tag{4.5}$$

Similarly, the bundle defined over $\bigcup_{a \epsilon A} U_a$

$$S^- : |\xi| \le \mu |\eta| \tag{4.6}$$

is mapped into itself under $d\phi^{-1}$, i.e.,

$$d\phi^{-1}(S^-) \subset S^- \ ; \tag{4.7}$$

moreover, if $(\xi_1, \eta_1) \in S^-$ and $(\xi_0, \eta_0)$ its pre-image then

$$|\eta_0| \geq \mu^{-1}|\eta_1| \; . \tag{4.8}$$

b) The above condition iii) expresses the instability of the mapping under iteration, as the horizontal components of a tangent vector gets amplified at least by $\mu^{-n}$ under $d\phi^n$ for $n \geq 1$, and the vertical component by $\mu^{-n}$ under $d\phi^{-n}$.

THEOREM 3.2. *If* $\phi$ *is a mapping which is continuously differentiable and satisfies the condition* i) *of the previous section and condition* iii), *just formulated, with* $0 < \mu < \frac{1}{2}$ *then condition* ii) *holds true with* $\nu = \mu(1-\mu)^{-1}$, *and hence the statements of Theorem* 3.1 *hold.*

*Proof.* It suffices to show that ii) is a consequence of i) and iii). Let $\gamma$ be a vertical curve in the strip $V_b(b \in A)$. As such it intersects every horizontal curve and, in particular, the boundaries of $U_a$. Thus $\hat{\gamma} = \gamma \cap U_a$ connects the horizontal boundaries of $U_a$, hence $\phi^{-1}(\hat{\gamma})$ connects the horizontal boundaries of $\phi^{-1}(U_a) = V_a$, i.e., $y = 0$ and $y = 1$. We want to show that $\phi^{-1}(\hat{\gamma}) = \phi^{-1}(\gamma) \cap V_a$ is a vertical curve (see Fig. 6). For this purpose we observe that $\gamma$ is a vertical curve, say

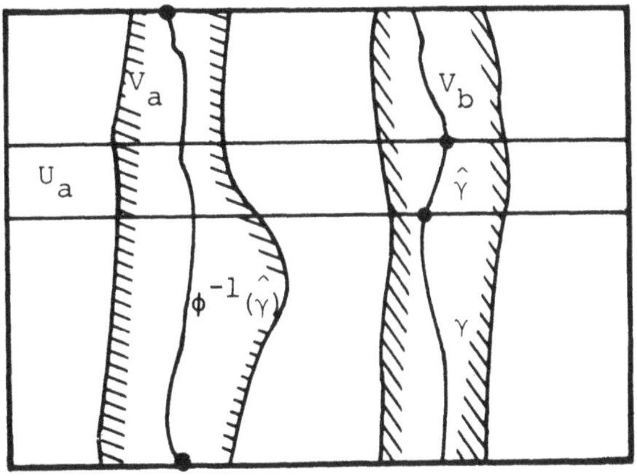


Fig. 6

$x = v(y)$, satisfying $|v(y_1) - v(y_2)| \le \mu |y_1 - y_2|$. Since $d\phi^{-1}$ maps $S^-$ into $S^-$ it follows by application of the mean value theorem that for any two points $(x_3, y_3), (x_4, y_4) \in \phi^{-1}(\hat{\gamma})$ one has

$$|x_3 - x_4| \le \mu |y_3 - y_4| .$$

This shows that $\phi^{-1}(\hat{\gamma})$ is the graph of a function $x = w(y)$ defined for $0 \le y \le 1$ satisfying

$$|w(y_3) - w(y_4)| \le \mu |y_3 - y_4| .$$

We apply this remark to the boundaries of $V \subset V_b$ and conclude that the pre-image of $\hat{V} = V \cap U_a$ hence $\phi^{-1}(V) \cap V_a = \phi^{-1}(\hat{V})$ is a vertical strip.

Secondly we verify the statement on the diameter for $0 < \mu < \frac{1}{2}$ and

$$\nu = \frac{\mu}{1-\mu} < 1 .$$

For this purpose let $p_1, p_2$ be two points on the two vertical boundaries of $\phi^{-1}(\hat{V})$, such that

$$d(\phi^{-1}(\hat{V})) = |p_1 - p_2|$$

and that $p_1, p_2$ have the same y-coordinates. Then the curve

$$p(t) = (1-t)p_1 + tp_2 \qquad (0 \le t \le 1)$$

is parallel to the x-axis and hence $\dot{p} \in S^+$. Therefore, the image curve

$$z(t) = \phi(p(t))$$

has its tangent

$$\dot{z} = d\phi\, \dot{p} \in S^+$$

by assumption iii). This shows that $z(0)$, $z(1)$ lie on a horizontal curve, namely the one obtained by extending $z(t)$ by horizontal segments. Thus, the points $(z(0), z(1))$ lie on a single horizontal curve, and two vertical lines at a distance $d(\hat{V})$. From (3.4) we conclude that

$$|z(0)-z(1)| \leq (1-\mu)^{-1} d(\hat{V}) \ .$$

Finally, writing $z(t) = (x(t), y(t))$ we have by the second assumption of iii) that

$$|\dot{x}| \geq \mu^{-1}|\dot{p}| > 0$$

hence $\dot{x}$ does not change sign and

$$|p_1 - p_2| = \int_0^1 |\dot{p}| dt \leq \mu \int_0^1 |\dot{x}| dt = \mu |x(1)-x(0)|$$
$$\leq \mu |z(1)-z(0)| \leq \mu(1-\mu)^{-1} d(\hat{V})$$

which verifies (3.7) with $\nu = \mu(1-\mu)^{-1}$.

c). *Hyperbolic structure of the invariant set*

In the case of a $C^1$-mapping one can, in addition, verify that the invariant set $I = \tau(S)$ is a hyperbolic set. That means, one can associate with every point $p \in I$ two linearly independent lines $L_p^+$, $L_p^-$ in the tangent space at $p$ such that $L_p^+$, $L_p^-$ vary continuously with $p \in I$, that the line bundles $L_p^{\pm}$ are invariant under $\phi$, i.e.,

$$d\phi \, L_p^{\pm} = L_{\phi(p)}^{\pm} \tag{4.9}$$

and that, with some constant $\lambda > 1$ and the norm $|\zeta| = \max(|\xi|, |\eta|)$ one has

$$\begin{cases} |d\phi(\zeta)| \geq \lambda|\zeta| & \text{for } \zeta \in L_p^+ \\ |d\phi^{-1}(\zeta)| \geq \lambda|\zeta| & \text{for } \zeta \in L_p^- \ . \end{cases} \tag{4.10}$$

We write the Jacobian $d\phi$ at a point $p$ as above with

$$\begin{pmatrix} f_x & f_y \\ g_x & g_y \end{pmatrix} = \begin{pmatrix} a & b \\ c & d \end{pmatrix}$$

and $\Delta = ad - bc$.

THEOREM 3.3. *If* $\Delta, \Delta^{-1} \leq \frac{1}{2}\mu^{-2}$ *then* I *is a hyperbolic set.*

*Proof.* We will construct the line bundle $L^+$ by the contraction principle. For this purpose we consider any line bundle in $S^+ = (S_p^+)$. We represent $L^+ = (L_p^+)$ in the form

$$\eta = \alpha_p \xi \quad \text{with} \quad |\alpha_p| \leq \mu$$

where $\alpha_p$ is a continuous function of $p \in I$. If we subject this bundle to the map $\phi$ we obtain

$$\eta_1 = \alpha^*_{\phi(p)} \xi_1$$

with

$$\alpha^*_{\phi(p)} = \frac{c + d\,\alpha_p}{a + b\,\alpha_p} \tag{4.11}$$

From our assumption (4.4) it follows that $\alpha^*_{\phi(p)}$ is defined for $|\alpha_p| \leq \mu$, and by (4.5) that $|a + b\alpha_p| \geq \mu^{-1}$. If $\beta_p$ defines a second line bundle, and $\beta_p^*$ its image we have

$$|\alpha^*_{\phi(p)} - \beta^*_{\phi(p)}| \leq \frac{|\Delta|}{|a + b\alpha_p|\,|a + b\beta_p|}\,|\alpha_p - \beta_p|$$

$$\leq \mu^2 |\Delta|\,|\alpha_p - \beta_p| \leq \frac{1}{2}|\alpha_p - \beta_p|$$

and since I is invariant under $\phi$

$$\sup_I |\alpha_p^* - \beta_p^*| \leq \frac{1}{2} \sup_I |\alpha_p - \beta_p| \ . \tag{4.12}$$

Thus if we define the mapping $\Phi$ by

$$(\Phi L)_p = d\phi L_{\phi^{-1}(p)} \tag{4.13}$$

for any line bundle $L_p$ in $S_p^+$ over I then $\Phi$ is a contraction in the norm $\sup_{p \in I} |\alpha_p - \beta_p|$. Hence, by the contraction principle, $S_p^+$ contains a unique continuous line bundle, denoted by $L_p^+$, invariant under $\Phi$, i.e., $\Phi L_p^+ = L_p^+$. Similarly, one constructs $L_p^-$ in $S_p^-$.

To verify the second property we observe that for $\eta = \alpha_p \xi$ with $|\alpha_p| \le \mu$ we have

$$|\xi_1| = |a + b\alpha_p| \, |\xi| \ge \mu^{-1} |\xi|$$

so that we can take $\lambda = \mu^{-1}$ in (4.10). With a similar argument for the vertical component Theorem 3.3 is proven.

d) Actually more than Theorem 3.3 can be proven. We recall that the points of I can be viewed as intersections of a horizontal curve U(s), defined by (3.14), and a vertical one V(s), see (3.12). Here U(s) is specified by $s_1, s_2, \ldots$ and V(s) by $s_0, s_{-1}, s_{-2}, \ldots$ . We claim

THEOREM 3.4. *Under the hypotheses* i) *and* iii) *with*

$$0 < \mu \le \frac{1}{2} \min \left( |\Delta|^{1/2}, |\Delta|^{-1/2} \right) \tag{4.14}$$

U(s), V(s) *are continuously differentiable curves whose tangents at points of* I *agree with the lines of the hyperbolic structure.*

*Proof.* It suffices to prove this for the horizontal curves U(s). We denote the set of all these horizontal curves by

$$\mathfrak{U} = \bigcup_s U(s)$$

and prove at first that

$$\phi^{-1}(\mathfrak{U}) \subset \mathfrak{U} \, . \tag{4.15}$$

Indeed, by (3.13) we have

$$U(s) = U_{s_1} \cap \phi(U(\sigma(s)))$$

hence, applying $\phi^{-1}$

$$\phi^{-1}(U(s)) = V_{s_1} \cap U(\sigma(s)) \subset U(\sigma(s))$$

which proves our contention (4.15).

From (4.15) it is clear that the map $\Phi$ defined by (4.13) is applicable to line bundles over $\mathfrak{U}$ and not only over I. In fact, the contraction property (4.12) holds also, if one replaces I by $\mathfrak{U}$, and the proof is clearly the same.

To prove our theorem we assign to any point $p \in \mathfrak{U}$ the set $T_p$ of limit lines of secants through two distinct points $q_1, q_2$ on the same horizontal curve U(s) as p in the limit $q_1, q_2 \to p$. More precisely, let $p = (x^*, y^*)$ lie on the horizontal curve $y = u(x)$ and let $T_p$ denote the set of lines $\eta = \alpha_p \xi$ with

$$\alpha_p = \lim_{\nu \to \infty} \frac{u(x_\nu) - u(x'_\nu)}{x_\nu - x'_\nu} \tag{4.16}$$

where $x_\nu \neq x'_\nu$ are two sequences tending to $x^*$ for which this limit exists. By (3.1) we have $|\alpha_p| \leq \mu$ and the set of $\alpha_p$ for a fixed p is closed. Let

$$\omega(T_p) = \max \alpha_p - \min \alpha_p \leq 2\mu$$

where the maximum and minimum is taken over the set (4.16). Thus $\omega(T_p) = 0$ implies that $y = u(x)$ has a tangent at p, and since we admitted two arbitrary sequences in (4.16), $\omega(T_p) = 0$ for all $p \in U(s)$ is equivalent to the statement that U(s) is continuously differentiable.

We observe that $T = (T_p)$ over $\mathfrak{U}$ is invariant under the mapping $\Phi$, defined in (4.13), i.e.,

$$\Phi T_p = T_p ,$$

as is seen from the mean value theorem. By the contraction property we have

$$\omega(T_p) = \omega(\Phi T_p) \leq \frac{1}{2} \omega(T_p) ,$$

hence, $\omega(T_p) = 0$ which proves that the horizontal curves are in $C^1$ and $T_p$ consists of a single line, the tangent line of U(s) at p. Since $T_p$ is invariant under $\Phi$, as is the line bundle $L_p^+$ $(p \in I)$ of the hyperbolic structure, the two coincide which proves Theorem 3.4.

Incidentally, if $\phi$ is $C^\infty$ so are the curves U(s), V(s), and if $\phi$ is real analytic, so are U(s), V(s). It is remarkable, that this holds without any smoothness assumptions on the boundary curves $U_a$, $V_a$.

e) *Remarks*

It will be our goal to verify the conditions i) and iii) for a mapping related to a given system of differential equations, in particular, the three-body problem. In this connection it is important to observe that the conditions have to be verified for $\phi$ only and not its iterates, while the Theorem 3.1 has implications for all iterates of $\phi$.

Closely related to this is the remark that the above homeomorphism $\tau$ depends continuously on the diffeomorphism $\phi$, if those diffeomorphisms are considered in the $C^1$-topology. This is a special case of a result of Anosov on the structural stability of so-called U-systems, diffeomorphisms of a compact manifold onto itself possessing a hyperbolic structure over the entire manifold and not only on a subset.

## 5. *The restricted three-body problem*

a) *Formulation*

We come to the application of the above results to the restricted three-body problem. The configuration, studied by Sitnikov, is the following: We have two mass points of equal mass $m_1 = m_2 > 0$ moving under Newton's law of attraction in the elliptic orbits while the center of mass is at rest. We consider a third mass point, of mass $m_3 = 0$, moving on the line L perpendicular to the plane of motion s of the first two mass points and going through the center of mass. Since $m_3 = 0$ the motion of the first two mass points is not affected by the third and from the symmetry of the situation it is clear that the third mass point will remain on the line L (Fig. 7). The problem is to describe the motion of the third mass point which is periodically excited by the first two.

We normalize the time so that the period of the primaries is $2\pi$, the mass unit so that the total mass is one, i.e., $m_1 = m_2 = \frac{1}{2}$ and the length

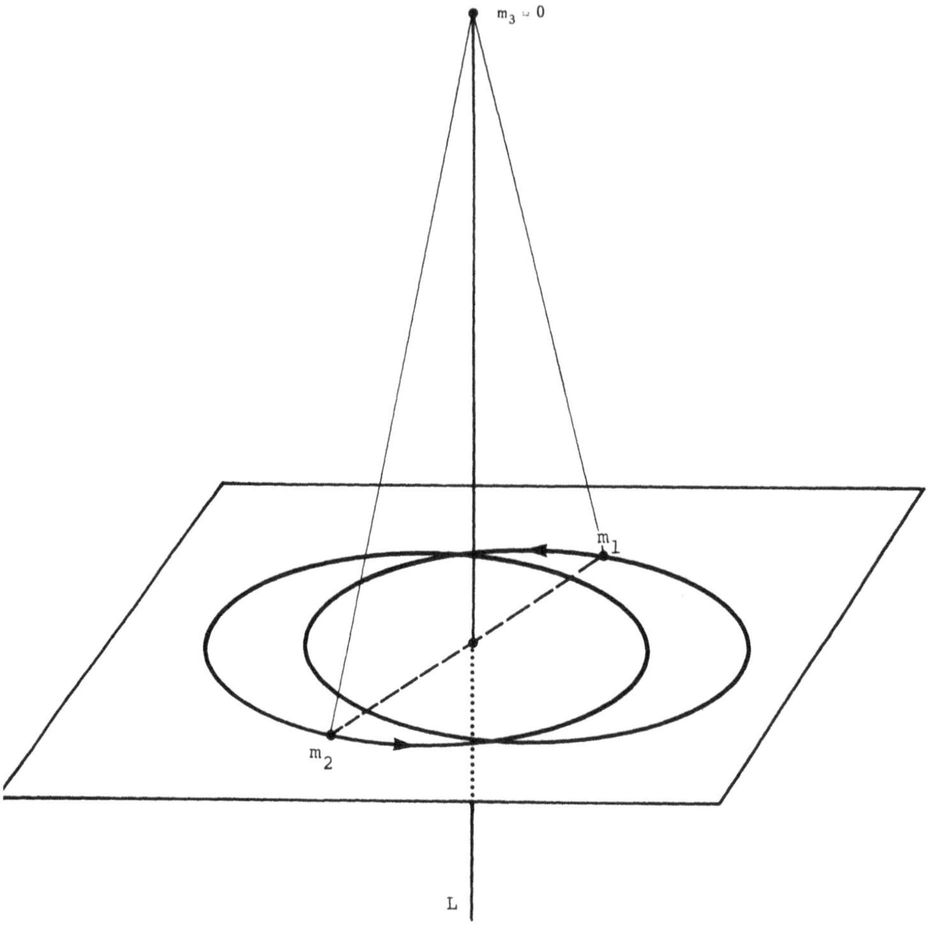


Fig. 7

unit so that the gravitational constant is one. Let $z$ be the coordinate describing the third mass point on $L$, so that $z = 0$ corresponds to the center of mass, and $t$ the time parameter. Then the differential equation takes the form

$$\frac{d^2z}{dt^2} = -\frac{z}{(z^2 + r^2(t))^{\frac{3}{2}}} \tag{5.1}$$

where $r(t) = r(t+2\pi) > 0$ is the distance of one of the primaries from the center of mass is an even function of $t$. With our normalization of units and denoting the eccentricity with $\varepsilon$ one finds for small value of $\varepsilon > 0$

$$r(t) = \frac{1}{2}(1 - \varepsilon \cos t) + O(\varepsilon^2) .$$

It is already a nontrivial matter to establish that all solutions with small $|z| + |\dot{z}|$ for $t = 0$ remain bounded for all real $t$. This is actually the case, and follows from the statements of Chapter II. On the other hand, if the velocity $\dot{z}(0)$ is very large the solution will tend to infinity. We will study the solutions near the critical escape velocity and show that there is a set of orbits to which one can associate a sequence of integers.

b) To formulate the result we consider a solution $z(t)$ with infinitely many zeroes $t_k (k = 0, \pm 1, \pm 2, \ldots)$ which are ordered according to size, $t_k < t_{k+1}$, $z(t_k) = 0$. Then we introduce the integers

$$s_k = \left[\frac{t_{k+1} - t_k}{2\pi}\right]$$

which measure the number of complete revolutions of the primaries between two zeroes of $z(t)$. This way we can associate to every such solution a doubly infinite sequence of integers. The main result can be expressed as the converse statement:

THEOREM 3.5. *Given a sufficiently small eccentricity* $\varepsilon > 0$ *there exists an integer* $m = m(\varepsilon)$ *such that any sequence* $s$ *with* $s_k \geq m$ *corresponds to a solution of the above differential equation.*

It seems remarkable that the $s_k$ can be chosen completely independently. If one chooses, for example an unbounded sequence the corresponding solution will be unbounded, but have infinitely many zeroes. Such solutions were discovered first by Sitnikov. But the above theorem allows also to finding infinitely many periodic orbits by choosing periodic sequences. Of course, to make this conclusion one has to know more about the correspondence mentioned in the theorem.

The above statement actually can be sharpened: First one can admit also half infinite sequences or finite sequences terminating with $\infty$ as we considered in the previous section. The corresponding solution, as expected, will escape after its last intersection with $z = 0$ if the sequence terminates on the right. Sequences terminating on the left correspond to capture orbit, which comes from $\infty$ and then perform infinitely many oscillations.

Secondly the smallness assumption on $\varepsilon$ is not really essential, as this statement holds true for all $0 < \varepsilon < 1$ except a discrete set of exceptional values. One could even admit $m_3$ to be small and positive as was observed by Alekseev.

Finally, we have to indicate the topological character of the correspondence between the sequences and the orbits. This will become clear from the geometrical considerations below.

c) *Reduction to a mapping*

We describe any solution $z(t) \not\equiv 0$ of (5.1) by giving its velocity $\dot{z}_0$ and time $t_0$ when $z(t_0) = 0$. We note that such a zero of $z(t)$ always exists. Otherwise we may assume that $z(t) > 0$ for all $t$, hence $\ddot{z} < 0$, i.e., $z(t)$ is concave for all real $t$ and positive which is impossible. If then $z(t_0) = 0$ one has $\dot{z}(t_0) \neq 0$ unless $z(t) \equiv 0$, by the uniqueness theorem.

Thus we can describe an arbitrary orbit by giving $t_0$ (mod $2\pi$) and $\dot{z}(t_0)$. Since the differential equation is invariant under the reflection $z \to -z$ we need not distinguish between positive and negative values of

$z$ and we set $v_0 = |\dot{z}(t_0)|$. The value $v_0 = 0$ corresponds to the trivial solution. Therefore, and since $t$ enters with period $2\pi$ into the differential equation we interpret $t_0$, $v_0$ as polar coordinates in a plane, $v_0$ being the radius and $t_0$ the angle variable.

Now we define a mapping $\phi$ on part of this plane by following a solution with initial conditions

$$z(t_0) = 0 \; ; \quad \dot{z}(t_0) = v_0$$

to its next zero, say $t_1 > t_0$, if it exists, and set $v_1 = |z(t_1)|$. Then the mapping $\phi$ takes $(v_0, t_0)$ into $(v_1, t_1)$. The question arises where $\phi$ is defined; we will prove in Chapter V:

LEMMA 1. *There exists a real analytic simple closed curve in* $R^2$ *in whose interior* $D_0$ *the mapping* $\phi$ *is defined. For* $(v_0, t_0)$ *outside* $D_0$ *the corresponding solutions escape.*

From the fact that the differential equations are invariant under the reflection $(z, t) \to (z, -t)$ one concludes easily: With the reflection

$$\rho : (v, t) \to (v, -t) \; ,$$

one has

$$\phi^{-1} = \rho^{-1}\phi\rho \; .$$

Moreover, one has

LEMMA 2. $\phi$ *maps* $D_0$ *onto a domain* $D_1 = \phi(D_0)$ *which agrees with the reflected domain* $D_1 = \rho(D_0)$. *Moreover,* $\phi$ *preserves the area element* vdvdt, *and*

$$\phi^{-1} = \rho^{-1}\phi\rho \; . \tag{5.2}$$

LEMMA 3. *If* $\varepsilon$ *is small enough and positive then* $D_0 \neq D_1$ *and the boundary curves* $\partial D_0$, $\partial D_1$ *intersect on the symmetry line nontangentially at a point* P (Fig. 8).

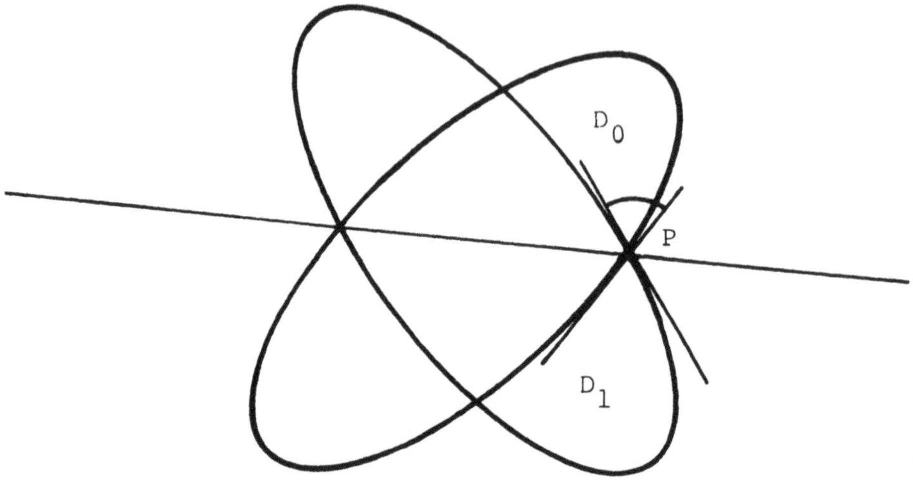


Fig. 8

This difference between the domain and range is the crucial property for the following. For $\varepsilon = 0$ this phenomenon does not occur, as both $D_0$ and $D_1$ are identical discs of radius 2 in that case.

d) Indeed for $\varepsilon = 0$ we have $r = \frac{1}{2}$ and the differential equation possesses the energy integral

$$\frac{1}{2}\dot{z}^2 - \frac{1}{\sqrt{z^2 + r^2}} = c \quad \text{where} \quad r = \frac{1}{2}$$

and c is independent of t and $\geq -2$. For $c < 0$ the above relation represents closed curves in the $z, \dot{z}$-plane, while for $c \geq 0$ these curves extend to infinity. For $c < 0$ the above curve intersects $z = 0$ at $\dot{z} = v$ with

$$v^2 = 2(c + \frac{1}{r}) = 2(c+2) < 4$$

so that the bounded solutions are represented by the disc $v < 2$.

To describe the mapping $\phi$ for $\varepsilon = 0$ observe, that by the above energy relation

$$v_1 = v_0 \quad \text{for} \quad v_0 < 2$$

and the time of return $t_1 - t_0 = T(v_0)$ is a function of $v_0$, and independent of $t_0$. Moreover, as $v_0 \to 2$ the time $T$ will approach infinity. Thus the mapping has the form

$$\begin{cases} v_1 = v_0 \\ t_1 = t_0 + T(v_0) \end{cases}$$

which means that concentric circles $v_0 = \text{const} < 2$ are rotated by an angle $T(v_0)$ which tends to $\infty$ as $v_0$ approaches 2. Thus the image of any radius is a curve spiralling infinitely about the origin as it approaches the boundary, see Fig. 9. A similar property holds for $\varepsilon > 0$, see Fig. 10.

LEMMA 4. *Let* $\gamma : v_0 = v_0(\lambda)$, $t_0 = t_0(\lambda)$ $(0 \le \lambda \le 1)$ *be a* $C^1$*-arc, such that* $\gamma$ *meets* $\partial D_0$ *in the endpoint* $p$ *corresponding to* $\lambda = 0$ *nontangentially, while* $\gamma$-$p$ *lies in* $D_0$. *Then the image curve*

$$\phi(\gamma) : v_1 = v_1(\lambda) \,, \quad t_1 = t_1(\lambda)$$

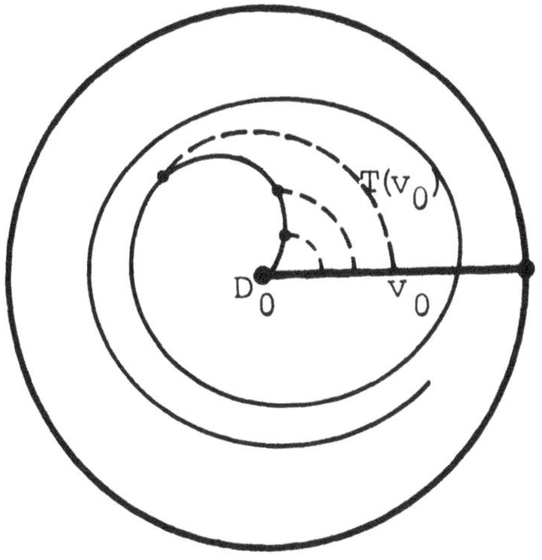


Fig. 9

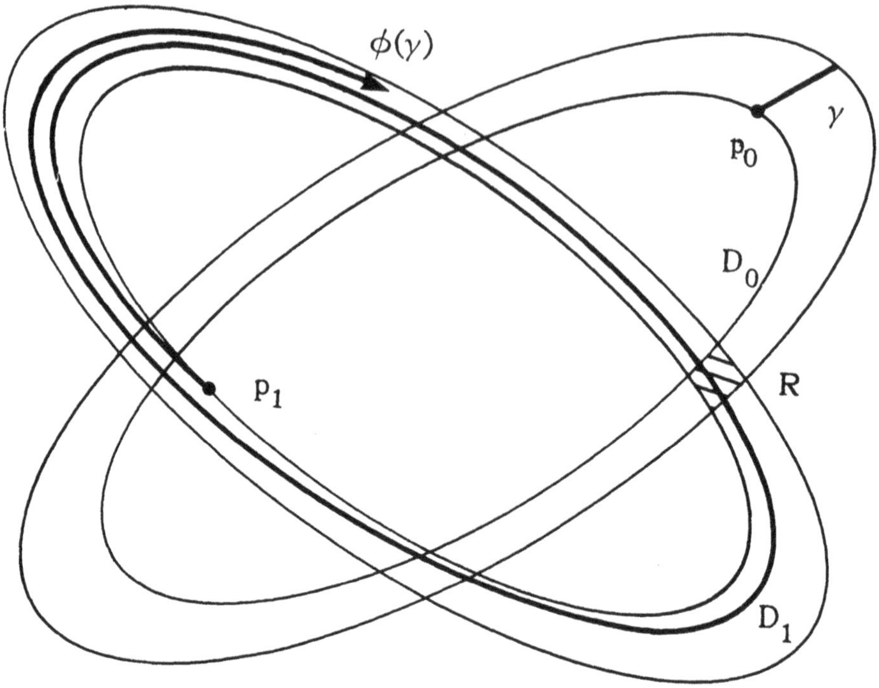


Fig. 10

*approaches the boundary* $\partial D_1$ *spiralling, i.e.,* $t_1(\lambda) \to \infty$ *as* $\lambda \to 0$ (Fig. 10).

Before relating this problem to the sequence shift of §4 we need one more technical lemma on the existence of a bundle of sectors near the boundary of $\partial D_0$. For $\delta > 0$, sufficiently small, we define $D_0(\delta)$ as the set of points in $D_0$ whose distance from $\partial D_0$ is less than $\delta$. Since $\partial D_0$ is continuously differentiable we can associate with any $p \in D_0(\delta)$ a unique closest point $q \in \partial D_0$, provided $\delta$ is small enough.

In $D_0(\delta)$ we define two bundles of sectors: The bundle $\Sigma_0 = \Sigma_0(\delta^{\frac{1}{3}})$ assigns to every point $p \in D_0(\delta)$ the set of lines which form an angle $\leq \delta^{\frac{1}{3}}$ with the line through p parallel to the tangent of $\partial D_0$ at q, the point on $\partial D_0$ closest to p. $\Sigma'_0$ assigns to every point the set of lines complementary to that of $\Sigma_0$ (Fig. 11). Similarly, $\Sigma_1$, $\Sigma'_1$ are the corresponding bundles of sectors over $D_1$, obtained, for example, by the reflection $\rho$ from $\Sigma_0$, $\Sigma'_0$.

LEMMA 5. *There exists a* $\beta$ *in* $0 < \beta < 1$, *such that for sufficiently small* $\delta$ *the mapping* $\phi$ *takes* $D_0(\delta)$ *into* $D_1(\delta^\beta)$ *and Jacobian mapping* $d\phi$ *of* $\phi$ *takes the bundle* $\Sigma'_0 = \Sigma'_0(\delta^{\frac{1}{3}})$ *into* $\Sigma_1 = \Sigma_1(\delta^{\frac{\beta}{3}})$.* *Moreover, if* $\zeta \in \Sigma'_0$, $\zeta_1 = d\phi(\zeta_0)$ *and* $\xi_0$ *the orthogonal projection of* $\zeta_0$ *into the center line of* $\Sigma'_0$ *and* $\xi_1$ *that of* $\zeta_1$ *into the center line of* $\Sigma_1$ *then*

$$|\xi_1| \leq \delta^{-\frac{1}{3}}|\xi_0| \ .$$

The situation is depicted in Fig. 12. We illustrate the force of this lemma by considering the image of a curve $\gamma$ in $D_0$ abutting on $\partial D_0$ like in Lemma 4. Since $\gamma$ is differentiable up to the boundary its tangent will lie in $\Sigma'_0$ for sufficiently small $\delta$. Hence $d\phi(\gamma)$ will lie in $\Sigma_1$, i.e., the direction of $d\phi(\gamma)$ deviates from the nearest tangent by at

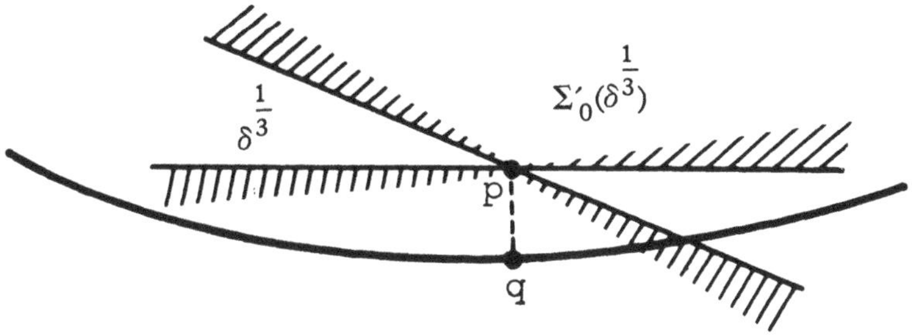


Fig. 11

* Since this lemma holds for all small $\delta$ we may consider the argument of $\Sigma'_0$, $\Sigma_1$ at point $p_\nu \in D_\nu (\nu = 1, 2)$ to be the third root of the distance of $p_\nu$ from $\partial D_\nu$.

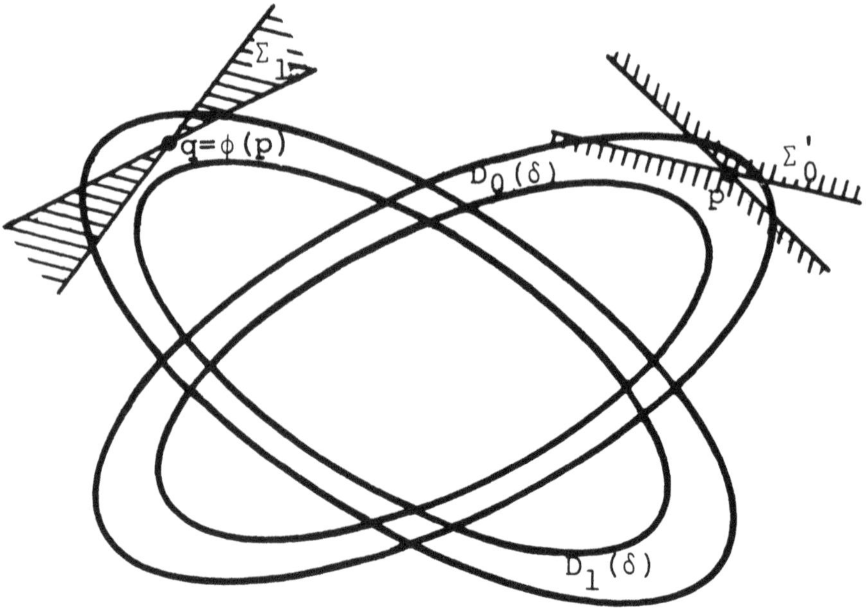


Fig. 12

most $\delta^{\frac{\beta}{3}}$. This shows that the image curve $\phi(\gamma)$ approaches $\partial D_0$ also with its tangent direction.

e) *The shift as subsystem of* $\phi$

The above lemmas will be proved in Chapter VI. With the aid of these lemmas we are in the position to prove

THEOREM 3.6. *The mapping* $\phi$ *in* $D_0$ *(defined for the restricted three-body problem in* §5c) *possesses the shift* $\overline{\sigma}$ *on* $D(\overline{\sigma}) \subset \overline{S}$ (see §3) *as a subsystem. Moreover, there exists a hyperbolic invariant set* I *homeomorphic to* S *on which* $\phi$ *is topologically equivalent to* $\sigma$.

To prove this result we consider the symmetric component R of the domain

$$D_0(\delta) \cap D_1(\delta)$$

containing in its closure the point P on the symmetry line. For sufficiently small $\delta$ this is a domain bounded by 4 differentiables curves which we refer to as sides. In fact, R will play the role of the square in §3 and we will verify the conditions i) and iii) of §3 and §4 from which the assertion follows.

For this purpose we observe that two of the sides of R are curves abutting on $\partial D_0$ nontangentially. Therefore, by Lemma 4, the image curves, and hence the image of R under $\phi$ spirals towards $\partial D_1$. Thus $\phi(R)$ and R intersect in infinitely many components, as is shown in Fig. 13. Aside from possibly finitely many of these components they will connect opposite sides of R. Dropping finitely many of the components of

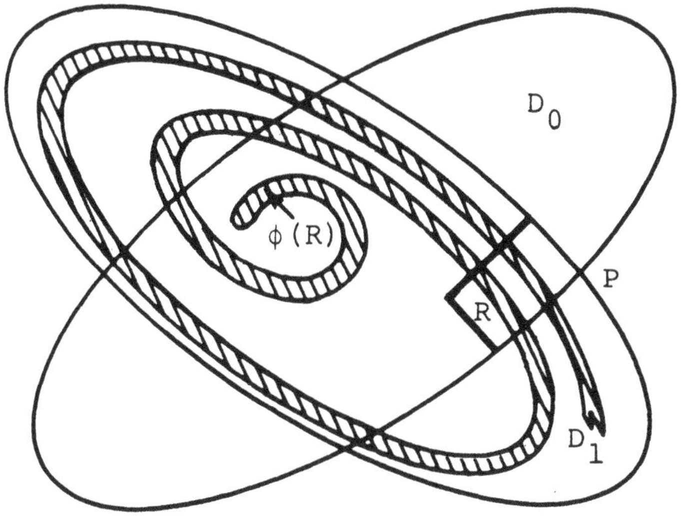


Fig. 13

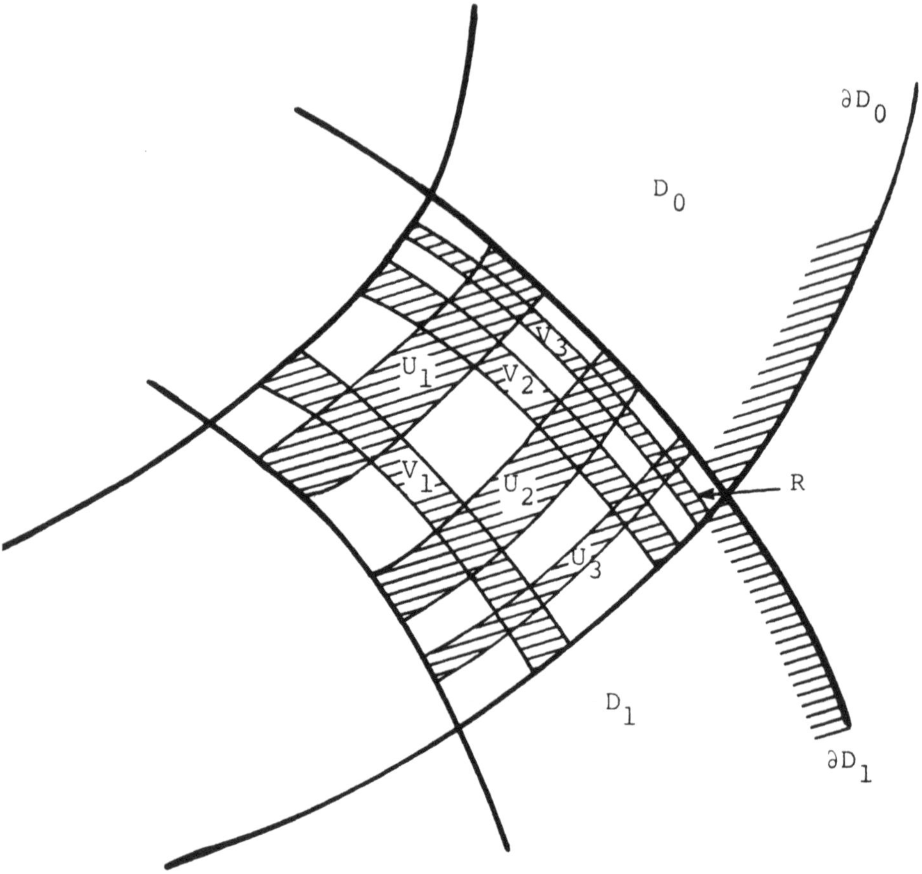


Fig. 14

$\phi(R) \cap R$ we will denote the others, in order, by $U_1, U_2, U_3, \ldots$ . Similarly, we set

$$V_k = \rho U_k \qquad (k = 1, 2, \ldots)$$

which are components of

$$\phi^{-1}(R) \cap R = \phi^{-1}\rho(R) \cap \rho R = \rho(\phi(R) \cap R) .$$

In fact, one shows that

$$\phi(V_k) = U_k \qquad (k = 1, 2, \ldots)$$

for which one may use (5.3) at the end of this section. By construction it is clear that $V_k (k = 1,\ldots)$ are disjoint closed sets, as are the $U_k$ (Fig. 14). To verify that $U_k$, $V_k$ satisfy the properties of horizontal and vertical strips in §3 we map R into the square Q. Since the sides of R are of length $\leq \delta$ one can achieve this with a mapping differing from a linear mapping by $O(\delta)$ in $C^1$-norm. To show that $V_k$ correspond to vertical strips it suffices to verify that the boundaries of $V_k$ interior to R have tangents whose direction is close to that of the tangent of $\partial D_1$ at P. This, in turn, is a direct consequence of Lemma 5, which implies that the angle between these directions is less than $\delta^{\frac{1}{3}} + O(\delta) \leq 2\delta^{\frac{1}{3}}$. Similarly, $U_k$ can be considered as horizontal strips as defined in §3.

Finally, we verify the condition iii) of §4. For this purpose we use the bundles $\Sigma_j$, $\Sigma'_j (j = 0,1)$ of Lemma 5 and note that over $R \subset D_0(\delta) \cap D_1(\delta)$ the sectors in $\Sigma_1(\delta^{\frac{1}{3}})$ are contained in $\Sigma'_0(\delta^{\frac{1}{3}})$, if $\delta > 0$ is small enough. In fact, by Lemma 3 the boundaries $\partial D_0$, $\partial D_1$ intersect at P transversally, and since the sectors in $\Sigma'_0$ contain all lines except those forming an angle $< \delta^{\frac{1}{3}} + O(\delta)$ with the tangent to $\partial D_0$ at P, they certainly contain the sectors of $\Sigma_1$, all of whose lines form an angle $\leq \delta^{\frac{1}{3}} + O(\delta)$ with the tangent of $\partial D_1$ at P. Thus $d\phi$, restricted to $R \cap \phi^{-1}(R)$, maps $\Sigma'_0$ into $\Sigma_1 \subset \Sigma'_0$. By the same token, we conclude that $d\phi$ maps $S^+ = \Sigma_1(\delta^{\frac{1}{3}})$ into itself. Indeed, for a point p in $\phi^{-1}(R) \cap R$ one has by Lemma 5 and the symmetry relation (5.2)

$$\phi^{-1}(p) \in \phi^{-1}(R) \subset \phi^{-1}(D_1(\delta)) \subset D_0(\delta^{\beta})$$

and so $\Sigma'_0(\delta^{\frac{\beta}{3}})$ is mapped into $\Sigma_1(\delta^{\frac{1}{3}})$, and in R, $\Sigma_1(\delta^{\frac{1}{3}}) \subset \Sigma'_0(\delta^{\frac{\beta}{3}})$, by the previous argument (Fig. 15). Hence applying $d\phi$ again, we observe that $d\phi$ maps $S^+ = \Sigma_1(\delta^{\frac{1}{3}})$ – considered over $\phi^{-1}(R) \cap R$ – into itself, as we wanted to show. Thus, $S^+$ restricted to

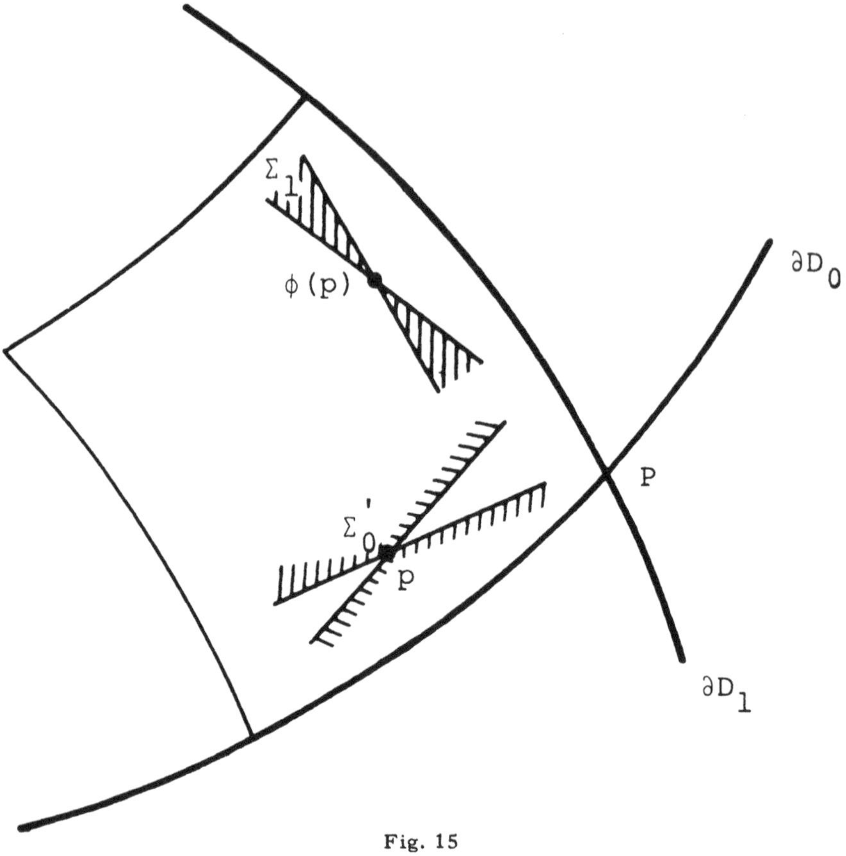


Fig. 15

$$\bigcup_{k=1}^{\infty} V_k \subset \phi^{-1}(R) \cap R$$

satisfies our hypothesis in iii). Similarly,

$$S^- = \Sigma_0(\delta^{\frac{1}{3}})$$

obtained from $S^+$ by reflection, restricted to $\bigcup_{k=1}^{\infty} U_k$ is mapped into itself by $d\phi^{-1}$. The remaining assertion on the stretching of the projections on the center line shows that the condition iii) of §4 is indeed satisfied with $\mu = c\delta^{\frac{1}{3}}$ where $c > 0$ depends on the angle under which $\partial D_+$ and $\partial D_-$ intersect at P.

This proves that Theorem 3.2 is applicable with $\mu = c\delta^{\frac{1}{3}}$ and the shift $\sigma$ on infinitely many symbols can be embedded as subsystem of $\phi$. Following §4 we can also obtain the shift $\bar{\sigma}$ of the compact sequence space $\bar{S}$ as subsystem of $\phi$, proving Theorem 3.6.

f) *Consequences*

In order to see that Theorem 3.5 is a consequence of Theorem 3.6 we have to observe merely, that the sets $V_k$ describe initial positions of those orbits for which the return time to $z = 0$ is of the form

$$t_1 - t_0 = 2\pi(k + c + \vartheta)$$

where c is a fixed constant dependent only on $\delta$ and $\vartheta$ lies in $0 \leq \vartheta < 1$. This is easily read off the geometrical description given above.

Thus we see that a sequence $s \in S$ corresponds to an orbit for which the integral part $[t_k - t_{k-1}/2\pi]$ of the return time is prescribed as an arbitrary sequence $s_k + [c]$. A sequence of the form

$$s = (\ldots s_{-1}, s_0, s_1, \ldots, s_{\lambda-1}, \infty)$$

corresponds to an orbit which for $t \to \infty$ escapes, and, in fact, for the corresponding point $p \in \tau(s)$ one has $\phi^{\lambda-1}(p) \in \partial D_0$, so that p is not in the domain of $\phi^{\lambda}$. If the sequence above is bounded for $k \leq 0$ then one can show easily that the corresponding solution $z(t)$ is bounded for $t \leq 0$. Thus we have found an orbit bounded for $t \leq 0$ but escaping for $t \to +\infty$, an escape orbit. The solution $z(-t)$ is, in this case, a capture orbit.

From an argument similar to that of Poincaré's recurrence theorem [8], [64], [88] one can show that the sets of orbits escaping for $t \to +\infty$ and for $t \to -\infty$ differ merely by a set of Lebesgue measure zero. Thus, in particular, the set of capture orbits is a set of measure zero, but, as the above result shows this set is *not empty* [87].

Also the existence of infinitely many hyperbolic periodic orbits follows at once, since any periodic sequence $s \in S$ corresponds to a periodic point $p = \tau(s)$ for $\phi$; thus $p$ is the initial point for a periodic orbit. Since $\tau(S)$ is a hyperbolic set all its periodic points are hyperbolic.

The existence of periodic orbits also follows easily from the following geometrical consideration: Let $\gamma$ be the arc of the symmetry line which lies in R (the diagonal of R through P). It is an arc abutting on $\partial D_0$ at P and therefore, by Lemma 4, the image $\phi(\gamma)$ spirals toward $\partial D_1$. Hence $\phi(\gamma)$ and $\gamma$ have infinitely many points $p_k$ of intersection. We order these points so that $p_1$ is the first point of intersection, as one traverses $\gamma$ from the endpoint in $D_0$, $p_2$ the second etc. It is clear that $\bigcup_{k=1}^{\infty} p_k = \gamma \cap \phi^{-1}(\gamma)$ invariant under $\phi$, since

$$\phi(\gamma \cap \phi^{-1}(\gamma)) = \phi(\gamma) \cap \gamma = \rho\phi^{-1}(\gamma) \cap \gamma = \phi^{-1}(\gamma) \cap \gamma$$

where we used $\phi = \rho\phi^{-1}\rho$ and $\rho\gamma = \gamma$. Considering the order of intersection it follows even that

$$\phi(p_k) = p_k \,, \tag{5.3}$$

i.e., the $p_k$ constitute infinitely many fixed points of $\phi$. Clearly, they correspond to constant sequences $s_a = m$ for $m$ large enough.

In his paper [33] Sitnikov established a class of solutions oscillating infinitely often unboundedly. For the two-body problem – or the present problem with $\varepsilon = 0$ – such solutions are clearly impossible, since there all oscillating orbits are bounded. The existence of unbounded oscillating solutions with infinitely many zeros follows readily by taking an unbounded sequence $s \in S$.

We observe that the mapping $\phi$ defined in c) admits the application of Theorem 2.11 which ensures the existence of invariant curves in $0 < v \leq r_0$ with $r_0 < 2$ if $|\varepsilon| < \varepsilon_0(r_0)$ is small enough. The corresponding solutions are then bounded for all real $t$ while the phenomenon of statistical behavior discussed in this section occurs near the critical velocity $v = 2$ where the perturbation effects are very large. On the other hand we will see in the next section, Theorem 3.9, that this phenomenon generally occurs also between the invariant curve in every neighborhood of an elliptic fixed point, like $v = 0$. Although this assertion is true generically for real analytic area-preserving mappings it has not been verified that the mapping $\phi$ associated with the restricted three-body problem belongs to this typical set of mappings.

## 6. *Homoclinic points*

a) The application to the restricted three-body problem may give the impression that the occurrence of the shift $\sigma$ as a subsystem is a relatively rare phenomenon. In fact, the opposite is the case as we want to show now. We will restrict ourselves to the discussion of diffeomorphisms in the plane since they are intimately related to differential equations in the three dimensional space.

Let $p, q$ be two different fixed points of such a diffeomorphism $\phi$ and $r \neq p, q$ a point such that $\phi^k(r)$ is defined for all integers $k$ and that

$$\phi^k(r) \to p \quad \text{for} \quad k \to -\infty$$

$$\phi^k(r) \to q \quad \text{for} \quad k \to +\infty \ .$$

In this case $r$ is called heteroclinic point. More generally, one calls a point heteroclinic, if $\phi^k(r)$ approaches one periodic orbit $\{p, \phi(p), \ldots, \phi^m(p) = p\}$ for $k \to -\infty$ and another $\{q, \phi(q), \ldots, \phi^m(q) = q\}$ for $k \to +\infty$. If these periodic orbits agree one speaks of a homoclinic point.

In case, $p, q$ are hyperbolic fixed points, i.e., $d\phi$ has real eigenvalues, $\lambda_1, \lambda_2$ satisfying $0 < |\lambda_1| < 1 < |\lambda_2|$ then the points approaching $p$ under iteration lie on an invariant curve $W_p^+$ through $p$, as do those escaping lie on a smooth curve, say $W_p^-$. In this case the heteroclinic points belonging to $p, q$ lie on the intersection $W_p^- \cap W_q^+ - \{p, q\}$. Usually, one requires that these curves intersect transversally.

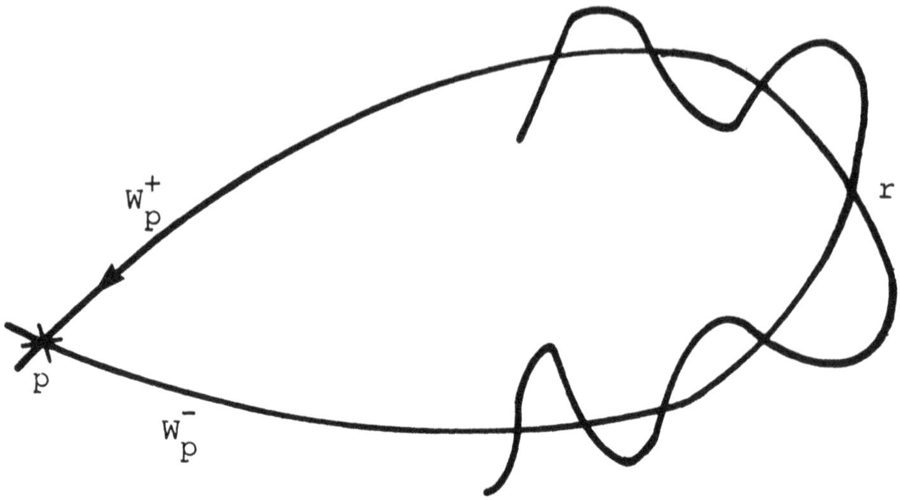


Fig. 16

In the following we will consider a homoclinic point belonging to a hyperbolic fixed point p of $\phi$ and such that $W_p^-$ and $W_p^+$ intersect transversally at some point $r \neq p$. Then r, as well as all points $\phi^k(r)$ are homoclinic points (Fig. 16).

The concept and the terminology of homoclinic and heteroclinic solutions goes back to H. Poincaré [90] who pointed to the complexity of the flow near such orbits (see No. 397). Also in the work of G. D. Birkhoff [89] the implications of this concept were studied. Here we want to stress that the presence of a homoclinic point belonging to a hyperbolic point p for which $W_p^+$ and $W_p^-$ intersect transversally at the point in question implies that the shift on infinitely many symbols is a subsystem of the mapping. This observation is due to Smale [80], at least as far as the embedding of a shift on *finitely* many symbols is concerned. The generalization to the shift on sequences of infinitely many symbols and of $\bar{\sigma}$ of §3 I learned from C. C. Conley. However, it has to be said, that these ideas, with less detail and rigor, were described already by Birkhoff in the above mentioned long publication [89] of 1935, where also the concept of sequences of symbols (see p. 184) can be found. In fact, our exposition follows quite closely that of Birkhoff, but we will supply the necessary estimates. One further reason for the choice of this exposition is that it has a very close similarity to that of the previous section. In fact, it turns out that the point P of intersection of $\partial D_0$, $\partial D_1$, the boundary curves of the domains $D_0$, $D_1$ of Lemma 2 in §5, plays the role of a homoclinic point, which belongs to an unstable periodic orbit at infinity. This is made more precise in Chapter VI, where it also turns out that this periodic orbit at infinity is not strictly hyperbolic but degenerate. Aside from such technical differences the geometrical situation is quite analogous to the one to be discussed near a homoclinic point.

b) *The transversal map* $\tilde{\phi}$

Near a homoclinic point r we construct a small quadrilateral R, two of its sides consisting of parts of $W_p^+$, $W_p^-$ and the others could be straight lines, parallel to the tangents of $W_p^+$, $W_p^-$ at r (see Fig. 17). For a point q we let $k = k(q)$ be the smallest positive integer for which $\phi^k(q) \in R$, if it exists. Denoting the set of $q \in R$, for which such a $k > 0$ exists by $D(\tilde{\phi})$ we set

$$\tilde{\phi}(q) = \phi^k(q) \quad \text{for} \quad q \in D(\tilde{\phi}) .$$

We will call $\tilde{\phi}$ the transversal map of $\phi$ for R. Of course, it has to be shown that $D(\tilde{\phi})$ is not empty. We will show more with the following theorem.

THEOREM 3.7. *If a* $C^\infty$*-diffeomorphism* $\phi$ *possesses a homoclinic point* r, *at which the curves* $W_p^+$, $W_p^-$ *of a hyperbolic fixed point* p *intersect transversally, then in any neighborhood of* r *the transversal map* $\tilde{\phi}$ *of a quadrilateral possesses an invariant subset* I *homeomorphic to the set* S *(space of sequences on* $N = \infty$ *symbols, defined in* §1b)) *via a homeomorphism* $\tau : S \to I$ *such that*

$$\tilde{\phi}\tau = \tau\sigma .$$

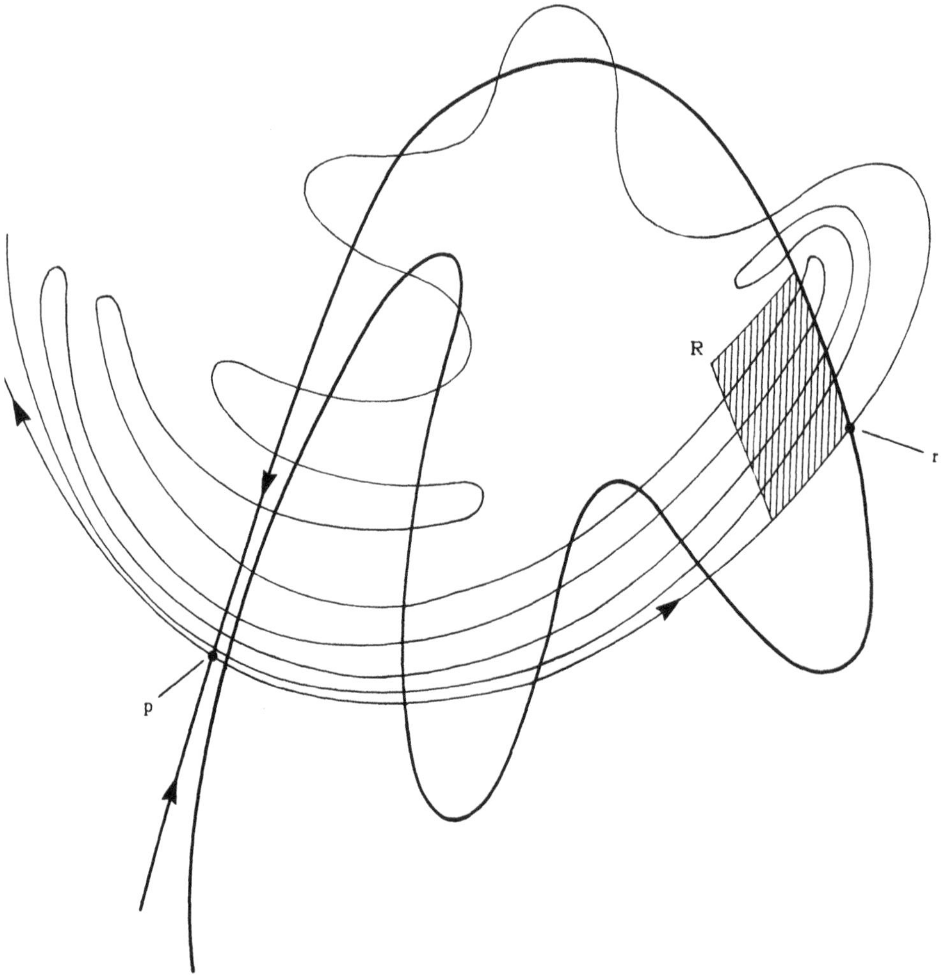


Fig. 17

*Moreover,* $\tau$ *can be extended to* $\bar{\tau}: \bar{S} \to \bar{I}$ *(notation §3) such that*

$$\tilde{\phi}\bar{\tau} = \bar{\tau}\bar{\sigma}$$

*if both sides are restricted to* $D(\bar{\sigma})$.

A very similar theorem, due to Smale asserts that in any neighborhood of a homoclinic orbit there exists an invariant set $I_2$ for some fixed power, say $\phi^\ell$, of $\phi$ such that $\phi^\ell$ on $I_2$ is homeomorphic to the sequence shift $\sigma$ on 2, or more generally on $N < \infty$ symbols. The minor difference of this from Theorem 3.7 is that $\tilde{\phi}$ is not given in terms of a fixed power of $\phi$, and on the other hand one can admit $N = \infty$ in Theorem 3.7.

The ideas of the proof of Theorem 3.7 are very similar to those for §5 by verifying the conditions i) and iii) of §3 and §4; see Chapter VI, §7. Here we want to point out that the above result implies the existence of many other homoclinic and heteroclinic points. Indeed, all sequences of $\bar{S}$ ending on both sides with the symbol $\infty$ (see §3) correspond to homoclinic points belonging to p. Since these sequences are dense in $\bar{S}$ we have

THEOREM 3.8. *With the assumptions of Theorem 3.7 the homoclinic points belonging to* p *are dense in* I.

This result indicates the complicated nature of the set of intersection of $W_p^+$ and $W_p^-$ which one may guess from the Fig. 17. Other homoclinic points are obtained by taking sequences for which the tail ends are periodic with the same block repeating at the right and the left. Heteroclinic orbits are obtained by sequences which possess two different repeating blocks for the right and for the left end. This shows that the existence of one single homoclinic point implies an infinity of such, and of heteroclinic points.

c) This leads us to the question of the existence of homoclinic points, which already was discussed by Poincaré [90], No. 399. First we remark that there is no difficulty in establishing such points for the diffeomorphism $\phi$ on the torus of §1c). Indeed, in that case $x = y = 0$ represents a hyperbolic fixed point and the straight lines along the eigenvectors define $W^+$ and $W^-$. Although these lines $W^+$, $W^-$ do not intersect in the plane, they do intersect transversally when viewed on the torus. This gives an example of a homoclinic point. In fact, using that every point with rational coordinates is a hyperbolic periodic point one can construct a dense set of homoclinic points on the torus.

However, the above example is very special, and undesirable since all its periodic points are unstable. Can homoclinic points exist also near elliptic fixed points? This is indeed the case, and, in fact, is the typical situation as the following theorem by Zehnder asserts. This theorem is formulated for real analytic mappings, $C^\infty$ or $C^\ell (\ell \geq 1)$ mappings. To formulate it we consider an area-preserving, real analytic mapping $\phi$ near an elliptic fixed point $x = y = 0$. Denoting the image point of $x, y$ by $x_1, y_1$ we write

$$x_1 = ax + by + \dots$$

$$y_1 = cx + dy + \dots$$

where the Jacobian has eigenvalues on $|\lambda| = 1$, $\lambda^2 \neq 1$. Using a device going back to Legendre we introduce $x, y_1$ as independent variables, writing

$$x_1 = d^{-1}(x + by_1) + \dots$$

$$y = d^{-1}(-cx + y_1) + \dots$$

provided that $d \neq 0$. One verifies easily that the area-preserving character of $\phi$ is tantamount to the requirement that the right-hand sides in the above formula are the derivatives of one single function, say $w(x, y_1)$, the so-called generating function, i.e.,

$$\begin{cases} x_1 = w_{y_1}(x, y_1) \\ y = w_x(x, y_1) \end{cases}, \quad w = (2d)^{-1}(-cx^2 + 2xy_1 + by_1^2) + \dots .$$

The advantage of this representation is that the coefficients of $w$ can be chosen freely, and the area-preserving property is taken care of. Consider the set $W$ of generating functions

$$w = \sum_{k+\ell \geq 2} c_{k\ell} x^k y_1^\ell \quad \text{with} \quad |c_{k\ell}| \leq 1 .$$

We introduce a topology in $W$ by using as neighborhoods of

$$w^* = \sum_{k+\ell \geq 2} c_{k\ell}^* x^k y_1^\ell$$

the set of $w \in W$ with $|c_{k\ell} - c_{k\ell}^*| < \varepsilon_{k\ell}$ where $\varepsilon_{k\ell}$ is an arbitrary positive sequence of numbers. A neighborhood in $W$ gives rise to a neighborhood of $\phi$ in the space of area-preserving mapping.

THEOREM 3.9 (Zehnder [91]). *If* $\phi$ *is an area-preserving real analytic mapping with* $(0,0)$ *as elliptic fixed points, then any neighborhood of* $\phi$ *contains an area-preserving mapping* $\psi$ *with the properties:* i) $(0,0)$ *is an elliptic fixed point with an eigenvalue not a root of unity,* ii) *every neighborhood of* $(0,0)$ *contains a homoclinic point.*

Actually Zehnder proves that the diffeomorphisms with an elliptic fixed point with the above properties form a residual set in the sense of Baire category, i.e., they are typical. We illustrate this situation with Fig. 18. Here F represents an elliptic fixed point and the simple closed curves are those described in Chapter II. In between the curves one finds periodic points like p, q in Fig. 18, fixed points of $\phi^3$. Here p is a hyperbolic point, for which $W_p^+$ and $W_p^-$ intersect transversally in r, a homoclinic point. According to the theorem one can embed the sequence

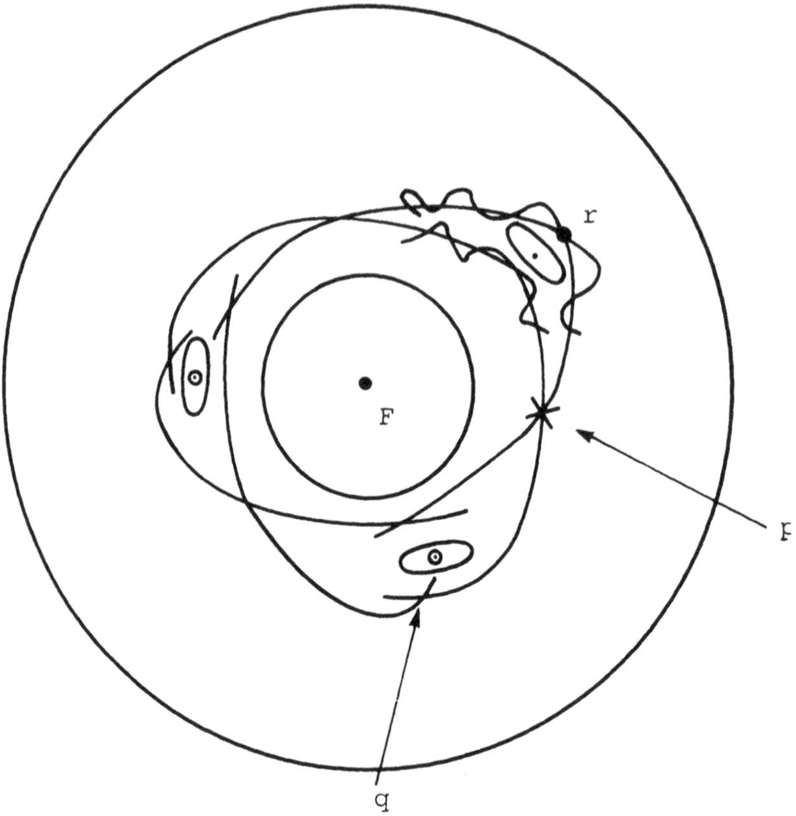


Fig. 18

shift $\sigma$ near r, and this complicated situation takes place in every neighborhood of F. On the other hand the elliptic periodic points, like q, are typically also surrounded by invariant curves, and also possesses homoclinic points nearby, in other words the situation of F is repeated near every such elliptic periodic point, giving rise to a hierarchy of invariant curves, periodic and homoclinic points.

d) *Nonexistence of integrals*

From a historical point of view the question of the existence of integrals for the three-body problem, and more generally, of a Hamiltonian system has been of great interest. We just mention the result of Brun, that any integral of the three-body problem, depending algebraically on the phase space variables is necessarily a function of the 10 known integrals. This remarkable theorem is, however, of little value for the dynamical description of the flow, since there may exist $C^1$ integrals or even real analytic integrals which are not algebraic. In the following we show that the results of §5 imply the nonexistence of a real analytic integral for the restricted three-body problem, and more generally, the nonexistence of such an integral near a homoclinic orbit.

For a diffeomorphism $\phi$ in a planar domain D we call a real-valued function f in D an integral of $\phi$ if f is not a constant, and $f(\phi(p)) = f(p)$. We will only consider real analytic functions f and show:

THEOREM 3.10. *Any diffeomorphism $\phi$ satisfying the conditions* i), iii) *of* §3 *and* §4 *and* (4.14) *does not possess a real analytic integral in* Q.

This theorem, which will be proven in Chapter VI, §8, implies that the transversal map $\tilde{\phi}$ of a diffeomorphism $\phi$ possessing a homoclinic point does not possess a real analytic integral. But if f were an integral for $\phi$ we conclude from $f(\phi^k(p)) = f(p)$ for $k = 1, 2, \ldots$ that it would be also an integral for $\tilde{\phi}$, showing that $\phi$ cannot possess such an integral.

In the same way it follows from §5 that the restricted three-body problem does not possess a real analytic integral, a remark pointed out by Alekseev [34], III, Corollary 14.

## CHAPTER IV

## FINAL REMARKS

a) In conclusion we see that the statistical behavior, as well as the stable behavior described by the quasi-periodic motion can take place side by side, even for analytic systems of differential equations. This is, of course, only the first step to clarifying the situation. We want to point out that the invariant sets constructed in Chapter III are all of measure zero in the natural measure of the embedding phase space and therefore can be considered negligible. On the other hand there are reasons to believe that there are invariant sets – say near a generic elliptic fixed point of an area-preserving mapping in the plane – of positive measure in which the mappings are ergodic. As far as I know there is no example of this type available. Ultimately it would be desirable to have some estimate for the ratio of the region on which the orbit shows a stable behavior to that in which statistical behavior prevails. These seem to be very difficult questions and even simplified model problems would be of interest, for which these questions could be decided.

In this connection it is worth mentioning some interesting results due to Anosov and Katok [69]. They constructed examples of measure-preserving $C^{\infty}$-diffeomorphisms and flows on smooth compact manifolds. For example, one might wonder whether the topological nature of the manifold has anything to do with the possibility of constructing an ergodic flow on a manifold. In 2 dimensions this is indeed the case, as the two-dimensional sphere, the projective plane and the Klein bottle do not admit ergodic flows. This is essentially a consequence of the Poincaré-Bendixson theorem. However, as a by-product of their investigation Anosov and Katok were able to construct smooth ergodic flows on any smooth compact manifold without boundary, provided its dimension is $\geq 3$. Although the ergodicity of these flows is destroyed under perturbation they provide new models of ergodic differential equations. For instance, with these methods Katok succeeded in constructing a smooth Hamiltonian of the form

$$H = \frac{1}{2} \sum_{k=1}^{n} (x_k^2 + y_k^2) + h(x, y)$$

where h and its higher derivatives are small, such that the flow on the energy surface $H = c$ for some small positive c is ergodic. This seems to be in contrast to the result on the existence of invariant tori of Chapter II. However, in this case the nondegeneracy condition on the Hessian is not satisfied, and this example shows that this condition is indeed necessary for those results.

b) All the above results refer to systems of finitely many degrees of freedom and the question arises whether or not there are some extensions to infinitely many degrees of freedom, or to partial differential equations. For example, one may ask for the stability of the trivial solution for the nonlinear hyperbolic differential equation

$$u_{tt} - u_{xx} = \alpha(u_x^3)_x$$

for $0 \leq x \leq 1$ with the boundary conditions $u = 0$ for $x = 0, 1$. If one approximates this differential equation by discretizing the interval $0 \leq x \leq 1$ one is led to a system of ordinary differential equations of the form

$$\ddot{q}_k = q_{k+1} - 2q_k + q_{k-1} + \alpha\{(q_{k+1} - q_k)^3 - (q_k - q_{k-1})^3\}$$

for $k = 1, 2, \ldots, N-1$ with $q_0 = q_N = 0$. This is a Hamiltonian system with the Hamiltonian

$$H = \frac{1}{2} \sum_{k=1}^{N} \{p_k^2 + (q_{k+1} - q_k)^2 + \frac{\alpha}{2} (q_{k+1} - q_k)^4\}$$

which falls into the category discussed in Chapter II. T. Nishida [92] discussed this problem and showed that at least, if $N$ is a power of 2 or a prime number the eigenvalues are rationally independent purely imaginary numbers, that the Birkhoff normal form up to order 4 is nondegenerate, so that for small positive $c$ the majority of the solution in $H \leq c$ are quasi-periodic, and thus behave like decoupled oscillators. This question had been studied numerically by Fermi, Pasta and Ulam in 1955 who expected that in the course of time different frequencies would be excited on account of the nonlinearity. The numerical results showed that this is in fact not the case, and the Theorem 2.8 of Chapter II can be viewed as a proof of this fact. However, the smallness condition on $c$ may become prohibitively severe as $N$ gets large, and the result may be actually insignificant without practical estimates.

One might expect that the solutions of the above partial differential equations behave in a similar way, since they are approximated by the second system. However, one can show (Lax [93]) that solutions can be continued in general only for a finite time. Thus the limits $N \to \infty$ and $t \to \infty$ cannot be interchanged and one cannot expect that the result for finite degree of freedom has any significance for the partial differential equation.

Another partial differential equation worth mentioning in this connection is the so-called Kortweg-deVries equation

$$u_t + uu_x + u_{xxx} = 0$$

which has been studied extensively from a numerical point of view. Remarkably, the wave solutions interact in a very regular manner without any scattering effects. Recently Gardner [94] and Sakharov and Faddeev [95] showed, that, at least from a formal point of view, this equation can be viewed as an integrable Hamiltonian system of infinite degrees of freedom. This raises the question whether some results of the finitely many degrees of freedom can be carried over to such a problem. These questions seem rather speculative and it is rather unlikely that the above results have any importance for systems of infinitely many degrees of freedom, but one rather should expect new phenomena which may require the development of new methods.

# CHAPTER V

# EXISTENCE PROOF IN THE PRESENCE OF SMALL DIVISORS

## 1. *Reformulation of Theorem* 2.9

### a) *Outline of chapter*

In this chapter we will discuss a convergence proof showing how one can cope with the small divisor difficulty, and carry out the complete proof of Theorem 2.9. The basic idea is to devise an iteration technique which converges so fast that the cumulative effect of the small divisors can be controlled. An example of such a rapidly converging iteration is provided by Newton's method, which, however, is not directly applicable to our problem. In Section 2 we will illustrate the new technique to be used on the simpler problem of finding a root of a function. In Section 3 we give the complete proof of Theorem 2.9. Section 4 contains indications of some generalizations of these results and in the appendix we give further details and improvements concerning the technique of Section 2 for finding a root.

The technique to be discussed has been applied to problems in other fields and it is desirable to formulate the resulting existence theorem in a general abstract form. Such a formulation can be found in L. Nirenberg [96] where a generalization of the Cauchy-Kowalevski existence theorem for partial differential equations is proven. Another area of applications of these techniques is the subject of stability and conjugacy of differentiable maps. We refer to the paper [97] by Sergeraert. However, the abstract existence theorems formulated and proven in these papers do not apply to our problem and therefore we will simply discuss the method for the problem at hand.

b) *Reversible systems, transformation properties*

Let $x = (x_1, \ldots, x_n)$, $y = (y_1, \ldots, y_n)$ denote two n-vectors and consider a system of differential equations

$$\dot{x} = f(x, y)$$
$$\dot{y} = g(x, y) \tag{1.1}$$

defined in a domain which is preserved under the reflection

$$\rho : (x, y) \to (-x, y) . \tag{1.2}$$

We consider systems which are reversible with respect to this reflection $\rho$, together with $t \to -t$ i.e., satisfy

$$f(-x, y) = f(x, y) , \quad g(-x, y) = -g(x, y) . \tag{1.3}$$

Under general coordinate transformations

$$x = u(\xi, \eta)$$
$$y = v(\xi, \eta) \tag{1.4}$$

where $\xi = (\xi_1, \ldots, \xi_n)$, $\eta = (\eta_1, \ldots, \eta_n)$ are the new variables, the simple form (1.3) of the reversibility is destroyed. But if we require that (1.4) commutes with $\rho$, i.e., satisfies

$$u(-\xi, \eta) = -u(\xi, \eta) , \quad v(-\xi, \eta) = v(\xi, \eta) \tag{1.5}$$

then the conditions (1.3) have the same form in the new coordinates.

To come to the reformulation of Theorem 2.9 we assume that $f, g$ in (1.1) are real analytic functions of $x$, $y$ and of period $2\pi$ in $x_1, x_2, \ldots, x_n$. These vector functions $f$, $g$ are defined in the real domain $T^n \times D$ where $D$ is an open domain in $R^n$ and $T^n$ stands for the n-dimensional torus. Since $f$, $g$ are real analytic, they are defined in a complex domain $X \times Y$ where $X$ is a complex neighborhood of $T^n$, invariant under $x_k \to x_k + 2\pi$ $(k = 1, 2, \ldots, n)$ and under the reflection $x \to -x$ and $Y$ is a complex neighborhood of $D$.

To preserve these additional properties of (1.1) under coordinate transformation we will require that

$$u(\xi,\eta) - \xi\ , \quad v(\xi,\eta) \tag{1.6}$$

have the period $2\pi$ in $\xi_1,\ldots,\xi_n$ and are real analytic in all variables.

This leads us to introduce the local group $\mathfrak{G}$ of coordinate transformations (1.4) given by real analytic functions u, v for which (1.6) are periodic in $\xi_1,\ldots,\xi_n$ and (1.5) holds; moreover, the transformations are required to be close to the identity transformation. Then a composition of two such transformations is defined in some subdomain and has the same properties there. Also the inverse of such a transformation exists in some subdomain and has the same properties.

This local group $\mathfrak{G}$ will play an essential role in the following proof. In the case one operates with Hamiltonian systems $\mathfrak{G}$ is replaced by the local group of canonical transformations near the identity. We chose to present the proof of Theorem 2.9 and not that of Theorem 2.8 because it is somewhat easier to work with transformations in $\mathfrak{G}$ than with canonical transformations; this is merely a technical difference, none in principle.

c) *Statement of Theorem* 5.1

Theorem 2.9 was formulated in terms of a small parameter $\mu$ while in the following we will express the theorem by giving a neighborhood in the class of differential equations for which the statement is valid. Furthermore, for $\mu = 0$ the unperturbed system took the form

$$\begin{aligned} \dot{x} &= F(y) \\ \dot{y} &= 0 \end{aligned} \tag{1.7}$$

where

$$\det F_y \neq 0\ . \tag{1.8}$$

In the following theorem we will take

$$F(y) = \omega + y \tag{1.9}$$

where $\omega = (\omega_1, \omega_2, \ldots, \omega_n)$ satisfy a condition of the form

$$|(j, \omega)| \geq \gamma |j|^{-\tau} \tag{1.10}$$

for all integer vectors $j \neq 0$. This function F certainly satisfies (1.8) and we will see later that the general case can be reduced to this case (1.9), for which $y = 0$, $x = \omega t + x(0)$ is a family of quasi-periodic solutions.

THEOREM 5.1 *Let* $0 < a < 1$ *and*

$$0 < \beta < \min\left(\frac{a}{4n + 4\tau + 1}, 1-a\right) \tag{1.11}$$

*and* $\gamma$, $\tau$ *as given by* (1.10). *Then there exists a* $\delta_0 = \delta_0(a, \beta, \gamma, \tau, n)$ *such that any system* (1.1) *satisfying*

$$|f - \omega - y| + |g| < \delta^{1+a} \quad \text{for} \quad |\text{Im}\, x_k| < \delta^{\beta}, \; |y_k| < \delta \; ^* \tag{1.12}$$

*for some* $\delta$ *in* $0 < \delta < \delta_0$ *as well as* (1.3) *can be transformed by a coordinate transformation of the form* (1.4) *in* $\mathfrak{G}$ *into a system*

$$\dot{\xi} = \phi(\xi, \eta)$$
$$\dot{\eta} = \psi(\xi, \eta)$$

*with*

$$\phi = \omega + \eta + O(\eta^2); \qquad \psi = O(\eta^2) . \tag{1.13}$$

*Moreover, the* u, v *in* (1.4) *are linear in* $\eta$ *and satisfy*

$$|u - \xi| + |v| < \delta \quad \text{in} \quad |\text{Im}\, \xi_k| < \frac{1}{2}\delta^{\beta}, \quad |\eta_k| < \frac{\delta}{2} . \tag{1.14}$$

---

* All estimates refer to complex neighborhoods.

The implication of (1.13) is that

$$\xi = \omega t + \xi(0) \;, \qquad \eta = 0$$

is a particular family of solutions, which via (1.4) yield the desired family of quasi-periodic solutions

$$x = u(\omega t + \xi(0), 0) \;, \qquad y = v(\omega t + \xi(0), 0)$$

of (1.1). Actually (1.13) contains more information about the derivatives of $\phi$, $\psi$ at $\eta = 0$ which is supplied by the proof.

We will postpone the proof of this simplified version of our theorem to §3 and show now that Theorem 5.1 implies Theorem 2.9. For this, we begin with the unperturbed system $\dot{x} = F(y)$, $\dot{y} = 0$ satisfying (1.7). It is defined in a domain $X \times Y$ where $Y$ is a complex neighborhood of an open domain $D$ in $R^n$. Hence, the image of $D$ under $F(y)$ is also open. Since the set of $\omega \in R^n$ satisfying (1.9) for some $\gamma > 0$ and any fixed $\tau > 0$ is dense in $R^n$ we can find a $\gamma$ and $\overset{0}{y} \subset D$ such that $\omega = F(\overset{0}{y})$ satisfies (1.10). In the following such a choice of $\omega$ will be fixed once and for all.

Since $F_y(\overset{0}{y}) = Q$ is a non-singular matrix we can introduce $z$ via

$$y = \overset{0}{y} + Q^{-1} z \tag{1.15}$$

and obtain for the unperturbed system of differential equations

$$\begin{cases} \dot{x} = F(\overset{0}{y} + Q^{-1} z) = \omega + z + G(z) \\ \dot{z} = 0 \end{cases}$$

where $G(z)$ vanishes quadratically at $z = 0$. Thus restricting $z$ to a domain

$$|z| < \delta$$

with sufficiently small $\delta > 0$ we can achieve that all $y$ in (1.15) with $|z| < \delta$ belong to $Y$ and such that

$$|G| < \text{const}\, \delta^2 < \frac{1}{2}\, \delta^{1+a} \quad \text{in} \quad |z| < \delta$$

since $\alpha < 1$. Finally, making the parameter $\mu$ in Theorem 2.9 sufficiently small we can achieve that the relevant system of that theorem satisfies the smallness condition (1.12) if $y$ is written in place of $z$ again.

The proof of Theorem 5.1 will show that $u$, $v$ depend analytically on any parameters on which $f$, $g$ depend analytically. Thus the functions $u(\xi, 0)$, $v(\xi, 0)$ and the quasi-periodic solutions defined by them and $\dot{\xi} = \omega$ also depend analytically on the parameter $\mu$, which implies Theorem 2.9.

d) *Linearized differential equations*

To interpret the additional information contained in (1.13) we introduce the concept of the linearized differential equations on an invariant manifold. For simplicity assume that the torus $y = 0$ is an invariant manifold of (1.1), i.e., that

$$g(x, 0) = 0 .$$

Then we associate with this manifold the linear system of differential equations

$$\begin{aligned}(\delta x)^{\cdot} &= f_x \delta x + f_y \delta y \\ (\delta y)^{\cdot} &= g_x \delta x + g_y \delta y\end{aligned}$$

where the arguments of the coefficients are given by any solution $(x(t), 0)$ of (1.1) on $y = 0$. Thus we may say that the linearized differential equation for $\dot{\xi} = \phi$, $\dot{\eta} = \psi$ on the invariant manifold $\eta = 0$ is given by

$$\begin{aligned}(\delta \xi)^{\cdot} &= \delta \eta \\ (\delta \eta)^{\cdot} &= 0 .\end{aligned}$$

In fact, it will be crucial in the approximation technique that the linearized equations of the transformed equations are controlled carefully.

We write down the transformation equations for the transformation (1.4) to take (1.1) into the new system. If $u_\xi$, $u_\eta$ denote the matrices of first derivatives of $u$ these equations are

$$(1.16)\qquad \begin{cases} u_\xi \phi + u_\eta \psi = f(u,v) \\ v_\xi \phi + v_\eta \psi = g(u,v) \,. \end{cases}$$

Since, by assumption $f - \omega - y$, $g$ are small we write

$$f = \omega + y + \hat{f} \,, \qquad g = \hat{g}$$

where $\hat{f}$, $\hat{g}$ indicate small functions. To set up the equation for $u(\xi,\eta)$, $v(\xi,\eta)$ we recall that they are supposed to be linear in $\eta$ and we set

$$(1.17)\qquad \begin{aligned} u &= \xi + \overset{0}{u} + \overset{1}{u}\eta \\ v &= \eta + \overset{0}{v} + \overset{1}{v}\eta \end{aligned}$$

where $\overset{0}{u}$, $\overset{0}{v}$ are n-vectors depending periodically on $\xi$ only, while $\overset{1}{u}$, $\overset{1}{v}$ are $n$ by $n$ matrices. Inserting (1.17) into (1.16) and setting $\eta = 0$ we obtain with (1.13)

$$(1.18)\qquad \begin{cases} \partial\overset{0}{u} - \overset{0}{v} = \hat{f}(\xi + \overset{0}{u}, \overset{0}{v}) \\ \partial\overset{0}{v} \qquad = \hat{g}(\xi + \overset{0}{u}, \overset{0}{v}) \end{cases}$$

and for the terms linear in $\eta$ we find

$$(1.19)\qquad \begin{cases} \partial\overset{1}{u} - \overset{1}{v} = -\overset{0}{u}_\xi + \hat{f}_x\overset{1}{u} + \hat{f}_y(I + \overset{1}{v}) \\ \partial\overset{1}{v} \qquad = -\overset{0}{v}_\xi + \hat{g}_x\overset{1}{u} + \hat{g}_y(I + \overset{1}{v}) \end{cases}$$

where

$$(1.20)\qquad \partial = \sum_{k=1}^{n} \omega_k \frac{\partial}{\partial \xi_k} \,.$$

The arguments of the derivatives of $\hat{f}$, $\hat{g}$ in (1.19) are, of course, $x = \xi + \overset{0}{u}$, $y = \overset{0}{v}$.

The proof of Theorem 5.1 requires then the solution of these nonlinear partial differential equations under the periodicity and symmetry requirements stated.

## 2. *Construction of the Root of a Function*

a) *Basic difficulty in solving* (1.18), (1.19)

We write the nonlinear equations (1.18), (1.19) in a symbolic way as

$$F(w) = 0 \tag{2.1}$$

where

$$w = \begin{pmatrix} u \\ v \end{pmatrix}$$

and

$$F(w) = Lw + \hat{F}(w)$$

is close to $Lw$ where $L$ is a linear operator, so that $L0 = 0$. The standard approach to solving such equations is to write (2.1) in the form

$$w = -L^{-1}\hat{F}(w) \tag{2.2}$$

and apply a contraction argument or an iteration method. This will fail here since $L^{-1}$ is an unbounded operator. Without writing $L$ explicitly at this point we note that inversion of $L$ involves solving the equation

$$\partial v(\xi) = g(\xi) \tag{2.3}$$

where $g(-\xi) = -g(\xi)$ and $v(-\xi) = v(\xi)$ are functions on the torus and $\partial$ is defined by (1.20). Expanding $g$ into a Fourier series

$$g(\xi) = \sum_j g_j e^{i(j,\xi)}$$

we have on account of the oddness $g_{-j} = -g_j$ and, in particular, $g_0 = 0$. Thus we obtain as solution

$$v = \sum_{j\neq 0} \frac{g_j}{i(j,\omega)} e^{i(j,\xi)} = \Lambda g \tag{2.4}$$

where $\Lambda g$ is defined by this sum. Under appropriate conditions we will verify the convergence of this series. This requires, in particular, a condition of the form (1.10) and it is clear that the $\omega_1, \omega_2, \ldots, \omega_n$ must be kept rationally independent.

Thus, if one thinks of solving (2.2) by an iteration process the unboundedness effect of $L^{-1}$ is accumulated and the convergence of the process is in doubt. To overcome this difficulty it is the aim to devise a scheme where the error $\varepsilon_n$ after $n$ steps is proportional to a power $\varepsilon_{n-1}^{\kappa}(\kappa > 1)$ of the $\varepsilon_{n-1}$. The unboundedness effect of $L^{-1}$ can then be controlled, as we will see.

The best known scheme having quadratic convergence $(\kappa = 2)$ is Newton's method, which takes the form

$$w_{n+1} = w_n - F'^{-1}(w_n)F(w_n)\,; \qquad w_0 = 0$$

where $F'(w_n)$ is the Frechet derivative of $F$ at $w = w_n$. This approach, however, is also doomed to fail, since $F'(w_n)$ is merely close to $L$, and it is not clear, whether the inverse of $F'(w_n)$ exists.

Thus we are led to search for an iteration method, which like Newton's method converges with an exponent $\kappa$ greater than $1$ but requires inversion of $L$ only. It is a priori not evident whether this is possible at all. To illuminate the idea of the following clearly we will illustrate the construction for the simple problem of finding a root of a function.

b) *Iteration scheme*

We consider a twice continuously differentiable real function $f(x)$ in the interval $|x - x_0| < r$. We assume $f'(x_0) \neq 0$. If then

(2.5) $$r^{-1}|f(x_0)| + r \sup_{|x|<r} |f''| < \delta|f'(x_0)|$$

is sufficiently small $f(x)$ will have a zero in $|x - x_0| < r$ which we will construct by a fast convergent iteration scheme (with $\kappa = \frac{3}{2}$), but not using the inverse of $f'$ except at $x = x_0$.

We may take $x_0 = 0$ and may assume that

(2.6) $$f(x) = x + \hat{f}(x)$$

is close to x near $x = 0$. The smallness condition (2.5) is equivalent to the smallness of

$$\|\hat{f}\|_r = \sup_{|x|<r} \{r^{-1}|\hat{f}(x)| + r|\hat{f}''(x)|\} \tag{2.7}$$

where the norm on the left is defined by this relation.

In the subsequent construction we will not only aim at the approximation of the root but of the inverse function, and this will be achieved by a succession of coordinate transformations. If $x = v(\xi)$ is an approximation to the inverse function of f then

$$\phi(\xi) = f(v(\xi)) \tag{2.8}$$

should differ less from $\xi$ than f(x) differs from x. Writing

$$v = \xi + \hat{v}\ , \quad \phi = \xi + \hat{\phi}$$

we have

$$\hat{\phi} = \hat{f}(v) + \hat{v}$$

and it is our aim to make $\hat{\phi}$ equal to zero, which requires the solution of $\hat{f}(v) + \hat{v} = 0$. Instead, we replace $\hat{f}(v)$ by $\hat{f}(\xi)$ and, furthermore, approximate $\hat{f}(\xi)$ by its linear part

$$\hat{f}_1(\xi) = \hat{f}(0) + \hat{f}'(0)\xi\ . \tag{2.9}$$

Thus we set

$$\hat{v}(\xi) = -\hat{f}_1(\xi) \tag{2.10}$$

and obtain

$$\hat{\phi}(\xi) = \hat{f}(\xi + \hat{v}) - \hat{f}_1(\xi) \tag{2.11}$$

That this gives rise to a useful approximation if $\|\hat{f}\|_r$ in (2.7) is small enough is the content of

LEMMA 5.1. *If* $\varepsilon = \|\hat{f}\|_r < \frac{1}{4}$ *and* $\rho$ *is chosen such that*

$$\varepsilon < \frac{\rho}{r} < 1 - 2\varepsilon$$

*then*

$$\|\hat{\phi}\|_\rho \leq \frac{c}{2}\left(\varepsilon \frac{\rho}{r} + \varepsilon^2 \frac{r}{\rho}\right) \quad \textit{with } c = 12 .$$

*With the optimal choice* $\rho = r\sqrt{\varepsilon}$ *we obtain the estimate*

$$\|\hat{\phi}\|_\rho \leq c\,\varepsilon^{\frac{3}{2}}$$

*in the much smaller interval* $|\xi| < \rho$.

*Proof*: By the definition (2.7) of $\|\hat{f}\|_r$ we have

$$|\hat{f}| \leq \varepsilon r, \quad |\hat{f}''| \leq \frac{\varepsilon}{r}$$

and, using the Taylor expansion we conclude that

$$|\hat{f} - \hat{f}_1| \leq \frac{1}{2} \sup |\hat{f}''| r^2 \leq \frac{1}{2} \varepsilon r$$

and hence, by (2.10)

$$|\hat{v}| = |\hat{f}_1| \leq \frac{3}{2} \varepsilon r < 2\varepsilon r \quad \text{in} \quad |x| < r . \tag{2.12}$$

Thus, by the assumption on $\rho$ we have for $|\xi| < \rho$ that

$$|\xi + \hat{v}(\xi)| < \rho + 2\varepsilon r < r$$

so that $x = \xi + \hat{v}(\xi)$ lies in $|x| < r$.

To estimate the derivatives of $\hat{v}$ we use that, for any function $g \in C''$ the derivative is dominated by $\|g\|_r$; in fact, one has

$$|g'| \leq 2\|g\|_r \quad \text{in} \quad |x| < r \tag{2.13}$$

as a simple consequence of the Taylor expansion: If $x \neq y$ lie in $(-r, +r)$ we have

$$g'(x) = \frac{g(y) - g(x)}{y-x} - \frac{1}{2} g''(y-x)$$

where $g''$ is taken at some intermediate point. Taking $y = x \pm r$ gives (2.13). Applying (2.13) to $g = \hat{f}$ and using (2.10) we have

$$|\hat{v}'| \leq 2\varepsilon \; . \tag{2.14}$$

To estimate $\|\hat{\phi}\|_\rho$ we use (2.11) and get for $|\xi| < \rho$

$$|\hat{\phi}(\xi)| \leq |\hat{f}(\xi + \hat{v}) - \hat{f}_1(\xi + \hat{v})| + |\hat{f}_1(\xi + \hat{v}) - \hat{f}_1(\xi)|$$

$$\leq \frac{1}{2} \sup_{|x|<r} |\hat{f}''|(\xi + \hat{v})^2 + |\hat{f}'(0)\hat{v}|$$

$$\leq \frac{1}{2} \frac{\varepsilon}{r} (\rho + \frac{3}{2} \varepsilon r)^2 + 2\varepsilon \frac{3}{2} \varepsilon r \; .$$

Using the assumption $\varepsilon r < \rho$ we get

$$|\hat{\phi}| \leq \frac{7}{2} \varepsilon \frac{\rho^2}{r} + 3\varepsilon^2 r \; .$$

To estimate the second derivative of $\phi$ we use (2.14) and get for $|\xi| < \rho$ and $\varepsilon < \frac{1}{4}$

$$|\hat{\phi}''| = |\hat{f}''(v)|(1 + \hat{v}')^2 \leq \frac{\varepsilon}{r}(1 + 2\varepsilon)^2 \leq \frac{5}{2} \frac{\varepsilon}{r} \; .$$

Combining the last inequalities we find

$$\|\hat{\phi}\|_\rho \leq 6\left(\varepsilon \frac{\rho}{r} + \varepsilon^2 \frac{r}{\rho}\right)$$

which proves the lemma.

Repeating this procedure we obtain the desired iteration method: For this purpose we call the given function $f_0 = f$ and the constructed approximation of its inverse $v_0 = v = \xi + \hat{v}$, and set $f_1 = f_0(v_0)$. This leads to a sequence of transformations $v_0, v_1, \ldots$ satisfying

$$f_{k+1} = f_k(v_k) \; .$$

With

$$u_k = v_0 \circ v_1 \circ \dots \circ v_{k-1}$$

we obtain

$$f(u_k(\xi)) = f_k(\xi) .$$

To make the above estimates applicable we define

$$\varepsilon_0 = \varepsilon , \quad \varepsilon_k = c\varepsilon_{k-1}^{\frac{3}{2}} \quad \text{for} \quad k = 1, 2, \dots$$

so that $\varepsilon_k$ converges rapidly to zero (exponent $\kappa = \frac{3}{2}$) if $0 < \varepsilon < c^{-2}$. Now we define the sequence $r_k$ by

$$r_0 = r , \quad r_k = r_{k-1}\varepsilon_{k-1}^{\frac{1}{2}} , \quad k = 1, 2, \dots .$$

Thus replacing $r$, $\rho$, $\varepsilon$ in Lemma 5.1 by $r_{k-1}$, $r_k$, $\varepsilon_{k-1}$ we obtain

$$\|f_k(x) - x\|_{r_k} \le c\varepsilon_{k-1}^{\frac{3}{2}} = \varepsilon_k \to 0$$

which by (2.13) implies that $f_k(0)$ and $f'_k(0) - 1$ both tend to zero. This cannot be said about the second derivative since in our norm it is multiplied by $r_k$ which tends to zero.

The convergence proof for $u_k$ is now straightforward. We use the estimates (2.12), (2.14)

$$r_k^{-1}|\hat{v}_k| , \quad |\hat{v}'_k| < 2\varepsilon_k \quad \text{for} \quad |\xi| < r_k$$

for $\hat{v}_k = v_k - \xi$. Noting that $\hat{v}'_k$ are constants the same is true for

$$u'_k = \prod_{\nu=0}^{k-1} (1 + \hat{v}'_\nu) .$$

For $k \to \infty$ this product converges since $\prod_{k=0}^{\infty} (1 + 2\varepsilon_k) = B$ converges for $\varepsilon < c^{-2}$. Thus $|u_k| \leq B$ for all $k \geq 0$ and from $u_{k+1} = u_k \circ v_k$ we conclude

$$|u_{k+1} - u_k| = |u_k \circ v_k - u_k| \leq B|\hat{v}_k| < 2B\varepsilon_k r_k$$

in $|\xi| < r_k$. Thus $u_k(0)$, $u'_k(0)$ both converge for $k \to \infty$, and hence $u_k(\xi)$, being linear, converges to a linear function $u(\xi)$ for all values of $\xi$.

Since

$$f(u(0)) = \lim_{k \to \infty} f(u_k(0)) = \lim_{k \to \infty} f_k(0) = 0$$

$$f'(u(0))u'(0) = \lim_{k \to \infty} f'(u_k(0))\, u'_k(0) = \lim_{k \to \infty} f'_k(0) = 1$$

it follows that $f(u(\xi))$ has a zero at $\xi = 0$ and its derivative is $= 1$ there. In other words $u(0) = \alpha$ is the desired root of $f(x)$ and $u'(0) = f'^{-1}(\alpha)$.

c) *Explicit recursion formulae*

The above derivation is certainly very impractical for finding a root, as it involves infinitely many coordinate changes. However, it is easy to make this scheme explicit and thus to obtain a useful method. For this purpose we have to recall that the $u_k(\xi)$ constructed above are linear functions and we set

$$u_k(\xi) = x_k + a_k \xi \ .$$

Then (2.10), (2.11) give rise to the recursion formulae

$$\text{(2.15)} \qquad \left\{ \begin{aligned} x_{k+1} &= x_k - a_k f(x_k) \\ a_{k+1} &= a_k + a_k(1 - f'(x_k) a_k) \end{aligned} \right\} \quad k = 0, 1, \ldots$$

with $x_0 = 0$, $a_0 = 1$. One observes that, indeed, $f'(x_k)$ never enters into the denominator. In fact, $a_k$ represents an approximation to $f'^{-1}(x_k)$ which is close enough to give rapid convergence. As a matter of fact, the convergence exponent of this method can be improved to

$$\kappa = \frac{1}{2}(1+\sqrt{5}) = 1.64 \ldots$$

which is the same as the rate for the "regula falsi" [98]. This will be shown in the appendix where we also give an improvement, due to O. H. Hald, of the method (2.15), by replacing $f'(x_k)$ in the second equation by $f'(x_{k+1})$. This actually leads to a convergence exponent $\kappa = 2$, and is quite useful for numerical computation.

In case $f(x) = x - Ax$ is a linear function the root is trivially zero; nevertheless (2.15) can be used to determine the inverse of $f' = 1 - A$ which is approximated by $a_k$. Since $x_k = 0$ the second line in (2.15)

$$a_{k+1} = a_k(2-(I-A)a_k) , \quad a_0 = 1$$

gives

$$a_{k+1} = (1+A)(1+A^2)\ldots(1+A^{2^k}) \quad \text{for} \quad k = 0,1,\ldots ,$$

converging to the well-known Euler product

$$(1-A)^{-1} = \prod_{k=0}^{\infty} (1+A^{2^k}) .$$

All these ideas can, of course, be generalized to finite dimensional spaces and to Banach spaces. But in the next section, where we present proof of Theorem 5.1, we will go back to the method of successive transformations of Section 2b) since it makes particularly appropriate use of transformation theory of differential equations.

## 3. *Proof of Theorem* 5.1

### a) *The approximating transformation*

For the proof of Theorem 5.1 we have to solve the equations (1.18), (1.19) for

(3.1) $$u = \xi + \overset{0}{u} + \overset{1}{u}\eta , \quad v = \eta + \overset{0}{v} + \overset{1}{v}\eta .$$

The procedure follows the idea of Section 2. We will solve (1.18), (1.19) only approximately and hoping to reduce the error after this transformation.

Then we will repeat this process successively and show that the composition of these transformations actually converges.

To define the approximating transformation we replace on the right-hand side $\hat{f}(\xi + \overset{0}{u}, \overset{0}{v})$ by $\hat{f}(\xi, 0)$ and $\hat{g}(\xi + \overset{0}{u}, \overset{0}{v})$ by $\hat{g}(\xi, 0)$. We replace the arguments on the right-hand side of (1.19) similarly and drop all products of $\overset{1}{u}$, $\overset{1}{v}$ and derivatives of $\hat{f}$, $\hat{g}$ since they would contribute quadratic error terms only. Thus we define u, v in (3.1) as solutions of

$$
(3.2) \qquad \begin{cases} \partial \overset{0}{u} - \overset{0}{v} = \hat{f}(\xi, 0) \\ \partial \overset{0}{v} \phantom{- \overset{0}{v}} = \hat{g}(\xi, 0) \end{cases}
$$

$$
(3.3) \qquad \begin{cases} \partial \overset{1}{u} - \overset{1}{v} = -\overset{0}{u}_{\xi} + \hat{f}_y(\xi, 0) \\ \partial \overset{1}{v} \phantom{- \overset{1}{v}} = -\overset{0}{v}_{\xi} + \hat{g}_y(\xi, 0) \end{cases}
$$

where $u - \xi$, $v$ must satisfy the periodicity condition as well as (1.5). Our first goal will be to solve this system of linear partial differential equations with constant coefficients.

b) *Solution of the linear equations*

We observe that the operator $\partial$ acts componentwise in the vector and matrix equations (3.2), (3.3) and it suffices to study the system of two scalar equations

$$
(3.4) \qquad \begin{cases} \partial \alpha - \beta = a(\xi) \\ \partial \beta \phantom{- \beta} = b(\xi) \end{cases}
$$

where a, b have period $2\pi$ in the $\xi_k$ and $a(-\xi) = a(\xi)$, $b(-\xi) = -b(\xi)$, according to the reversibility. We seek solutions of (3.4) which also are periodic in $\xi_k$ and satisfy

$$
(3.5) \qquad \alpha(-\xi) = -\alpha(\xi) \;, \quad \beta(-\xi) = \beta(\xi) \;.
$$

Using Fourier expansion and the notation (2.4) we see that

$$
\beta = \Lambda b + \beta_0
$$

since b, being odd, has no constant term in its Fourier expansion; $\beta_0$ is a constant, so far arbitrary, which will be used to balance the constant terms in (3.4). If $a_0$ is the constant term of a in its Fourier expansion we set $\beta_0 = -a_0$ and obtain

$$\begin{aligned} a &= \Lambda(a+\beta) = \Lambda((a-a_0) + \Lambda b) \\ &= \Lambda(a-a_0) + \Lambda^2 b \ . \end{aligned}$$

The solution so obtained satisfies (3.5), and is uniquely defined.

For the following it will be important to have estimates for these solutions. For this purpose we define for a real analytic function $a(\xi)$, defined in the complex domain $|\mathrm{Im}\ \xi_k| \le r$, the norm

$$|a|_r = \sup_{|\mathrm{Im}\ \xi_k|<r} |a(\xi)| \ .$$

We assume that b is real analytic in the same domain.

LEMMA 5.2. *If* $0 < \rho < r < 1$ *then the solution* $a$, $\beta$ *of* (3.4), (3.5) *is real analytic in* $|\mathrm{Im}\ \xi_k| < \rho$ *and satisfies*

$$|a|_\rho + |\beta|_\rho \le \frac{c_1}{(r-\rho)^\nu} (|a|_r + |b|_r) \tag{3.6}$$

*where* $c_1$ *is a positive constant depending on* $\gamma$, $\tau$, n *and* $\nu = 2(\tau + n)$.

The proof is straightforward: For the Fourier coefficients $a_j$ of $a = a(\xi)$ one obtains easily

$$|a_j| \le |a|_r e^{-|j|r} \ , \quad |j| = \sum_{k=1}^{n} |j_k|$$

which expresses the exponential decay of the Fourier coefficients of a real analytic function. Thus, using the inequalities (1.10) one can estimate the small divisor series $\Lambda(a-a_0)$ as follows:

$$|\Lambda(a-a_0)|_\rho \le \sum_{j\ne 0} |a_j|\, |(j,\omega)|^{-1} e^{|j|\rho}$$

$$\le |a|_r\, \gamma^{-1} \sum_{j\ne 0} |j|^\tau\, e^{-|j|(r-\rho)} .$$

It is easy to estimate this sum by

$$c_1(r-\rho)^{-\tau-n} .$$

Actually more refined estimates lead to such an inequality with the exponent $-\tau$ (see [65]). Recalling that $\alpha$ contained $\Lambda^2$ we obtain (3.6).

The inequality (3.6) gives rise to an estimate for the solution u, v of (3.2), (3.3), which we derive now. For this purpose we set

$$h(\xi,\eta) = \begin{pmatrix} f \\ g \end{pmatrix}, \quad \hat{h} = \begin{pmatrix} f-\omega-\eta \\ g \end{pmatrix} = \begin{pmatrix} \hat{f} \\ \hat{g} \end{pmatrix} \tag{3.7}$$

where $f = f(\xi,\eta)$, $g = g(\xi,\eta)$ are the vector functions of (1.1). We define the norm

$$|\hat{h}|_{r,s} = \sup |\hat{h}(\xi,\eta)|$$

where the supremum is taken over the complex domain

$$|\mathrm{Im}\, \xi_k| < r , \quad |\eta_k| < s ,$$

and all components of the vector $\hat{h}$. Finally, we denote by $(h)_1$ the linear part

$$(h)_1 = h(\xi,0) + h_\eta(\xi,0)\eta . \tag{3.8}$$

LEMMA 5.3. *One has*

$$|(h)_1|_{r,s} \le c_2 |h|_{r,s}$$

*where* $c_2 = (n+1)$.

*Proof*: By Cauchy's inequality one has

$$\left|\frac{\partial}{\partial \eta_k} h(\xi, 0)\right| \leq \frac{1}{s} |h|_{r,s}$$

hence

$$|(h)_1|_{r,s} \leq (1+n)|h|_{r,s} \ .$$

We set

$$(3.9) \qquad w = \begin{pmatrix} u \\ v \end{pmatrix} = \begin{pmatrix} \xi + \overset{0}{u} + \overset{1}{u}\eta \\ \eta + \overset{0}{v} + \overset{1}{v}\eta \end{pmatrix} ; \qquad \hat{w} = w - \begin{pmatrix} \xi \\ \eta \end{pmatrix} .$$

and estimate for the solution of (3.2), (3.3) the norm $|\hat{w}|_{\rho,s}$ in terms of $|\hat{h}|_{r,s}$. Using Lemma 5.2 we obtain from (3.2)

$$|\hat{w}(\xi, 0)|_{r',s} \leq \frac{c_1}{(r-r')^{\nu}} |\hat{h}(\xi, 0)|_{r,s}$$

for $0 < r' < r$. To estimate the first terms $w_\xi(\xi, 0)$ on the right-hand side of (3.3) we use Cauchy's inequality to get for $r'' < r' < r$

$$|\hat{w}_{\xi_k}(\xi, 0)|_{r'',s} \leq \frac{1}{r'-r''} |\hat{w}(\xi, 0)|_{r',s} \leq \frac{1}{r'-r''} \frac{c_1}{(r-r')^{\nu}} |\hat{h}(\xi, 0)|_{r,s}$$

Thus taking for example $r-r' = r'-r'' = \frac{1}{3}(r-\rho)$ we find

$$|\hat{w}_\xi(\xi, 0)|_{r'',s} \leq \frac{c_3}{(r-\rho)^{\nu+1}} |\hat{h}(\xi, 0)|_{r,s}$$

and applying Lemma 5.2 to (3.3) we obtain

$$(3.10) \qquad |\hat{w}|_{\rho,s} \leq \frac{c_4}{(r-\rho)^{\lambda}} |(\hat{h})_1|_{r,s} \leq \frac{c_5}{(r-\rho)^{\lambda}} |\hat{h}|_{r,s}$$

where

$$(3.11) \qquad \lambda = 2\nu + 1 = 4\tau + 4n + 1$$

and $c_5 = c_4 c_2$; here we used Lemma 5.3. This is the desired estimate for the approximate transformation (3.1) in the notation (3.9).

c) *Iteration process*

Now we transform (1.1) by the coordinate transformation $x = u(\xi, \eta)$, $y = v(\xi, \eta)$ defined by (3.1), (3.2), (3.3) and show that the transformed system

$$\dot{\xi} = \phi(\xi, \eta) = \omega + \eta + \hat{\phi} \,, \quad \dot{\eta} = \psi(\xi, \eta) = \hat{\psi}$$

has smaller error terms $\hat{\phi}$, $\hat{\psi}$ than (1.1), provided one restricts the domain properly.

For simplicity of the notation we combine the n-vectors $\xi$, $\eta$ to a 2n-vector $\zeta$, and $\phi$, $\psi$ to $X$, and write the new system in the form

$$\dot{\zeta} = X(\zeta) = \begin{pmatrix} \omega + \eta \\ 0 \end{pmatrix} + \hat{X} \tag{3.12}$$

LEMMA 5.4. *Let* $0 < \alpha < 1$, *and* $\beta$, $\lambda$ *like in* (1.11), (3.11) *respectively. We assume*

$$0 < \rho < r < 1 \,, \quad 0 < 2\sigma < s < (r-\rho)^{\frac{1}{1-\alpha}} < 1 \tag{3.13}$$

*and with the constant* $c_5$ *from* (3.10)

$$2 c_5 s^\alpha < (r-\rho)^\lambda \,. \tag{3.14}$$

*If the system* (1.1) *satisfies*

$$|\hat{h}|_{r,s} < s^{1+\alpha} \tag{3.15}$$

*in the notation* (3.7) *then the transformation constructed above takes* (1.1) *into* (3.12) *with*

$$|\hat{X}|_{\rho,\sigma} < c_6 s^{1+\alpha} \{ (r-\rho)^{-\lambda} s^\alpha + \left(\frac{\sigma}{s}\right)^2 \} \,. \tag{3.16}$$

*Moreover, in the notation* (3.9) *we have*

$$|\hat{w}|_{\rho,\sigma} < c_5 (r-\rho)^{-\lambda} s^{1+\alpha} \,. \tag{3.17}$$

We delay the proof of this lemma and show first how it leads to the proof of Theorem 5.1. For this purpose we choose $\sigma$ so that

$$2c_6(r-\rho)^{-\lambda}s^{\alpha} = \left(\frac{\sigma}{s}\right)^{1+\alpha}$$

or

$$\sigma = c_7(r-\rho)^{-\frac{\lambda}{1+\alpha}} s^{1+\frac{\alpha}{1+\alpha}}, \quad c_7 = (2c_6)^{\frac{1}{1+\alpha}} . \tag{3.18}$$

We reserve the freedom to make $c_5$ larger, which by (3.18) and (3.14) implies that $\frac{\sigma}{s}$ can be made small. Thus we see that (3.16) takes the form

$$|\hat{X}|_{\rho,\sigma} < s^{1+\alpha}\left(\frac{1}{2}\left(\frac{\sigma}{s}\right)^{1+\alpha} + c_6\left(\frac{\sigma}{s}\right)^{2}\right)$$

$$< s^{1+\alpha}\left(\frac{1}{2}\left(\frac{\sigma}{s}\right)^{1+\alpha} + \frac{1}{2}\left(\frac{\sigma}{s}\right)^{1+\alpha}\right) = \sigma^{1+\alpha}$$

provided $\sigma$ is chosen as in (3.18) and $c_5$ large enough. This estimate has the same form as (3.15) and is therefore well suited for an iteration process.

To carry out such an iteration we define two sequences $r_k$, $s_k$ by

$$\left\{\begin{aligned} r_k &= \left(\frac{1}{2} + \frac{1}{2^k}\right) r_0 , \\ s_{k+1} &= c_7(r_k - r_{k+1})^{-\frac{\lambda}{1+\alpha}} s_k^{1+\frac{\alpha}{1+\alpha}} , \end{aligned}\right. \qquad (k = 0, 1, \ldots) \tag{3.19}$$

where $r_0$, $s_0$ will be chosen presently.

From the recursion formula for $s_k$ we see that $s_k$ tends to zero if $s_0$ is small enough. In fact, (3.19) implies

$$s_{k+1} \leq c_8^{k+1} r_0^{-\frac{\lambda}{1+\alpha}} s_k^{1+\frac{\alpha}{1+\alpha}}$$

or

$$\left(r_0^{-\frac{\lambda}{\alpha}} s_{k+1}\right) \leq c_8^{k+1} \left(r_0^{-\frac{\lambda}{\alpha}} s_k\right)^{1+\frac{\alpha}{1+\alpha}}$$

and it follows that $s_k$ tends to zero if

$$r_0^{-\frac{\lambda}{\alpha}} s_0 \tag{3.20}$$

is small enough. This follows simply from the fact that the sequence $\varepsilon_k$ defined by

$$\varepsilon_{k+1} = c_8^{k+1} \varepsilon_k^{\kappa} , \quad \kappa > 1$$

converges to zero for sufficiently small positive $\varepsilon_0$. Indeed setting

$$\delta_k = c_8^{pk+q} \varepsilon_k , \quad p = \frac{1}{\kappa-1} , \quad q = \frac{\kappa}{(\kappa-1)^2}$$

the above recursion relation takes the form

$$\delta_{k+1} = \delta_k^{\kappa}$$

which converges to zero if $0 \leq \delta_0 < 1$, hence $\varepsilon_k \to 0$ if $\varepsilon_0 < c_8^{-q}$. Clearly the convergence has the exponent $\kappa = 1 + \frac{\alpha}{1+\alpha} < \frac{3}{2}$. We also observe that all conditions (3.13), (3.14) hold for $r = r_k$, $\sigma = r_{k+1}$, $s = s_k$, $\sigma = s_{k+1}$, $k = 0, 1, \ldots$ if (3.20) is chosen small enough.

To prove Theorem 5.1 we set

$$s_0 = \delta , \quad r_0 = \delta^{\beta} \tag{3.21}$$

and observe that on account of (1.11), (3.11) we have $\beta < \frac{\alpha}{\lambda}$, hence the quantity (3.20) can be made small by making $\delta$ small. With this choice of $r_0$, $s_0$ we apply the transformation constructed above and Lemma 5.4 with $r = r_0$, $s = s_0$, $\rho = r_1$, $\sigma = s_1$ and obtain a new system with a smaller error. We repeat this procedure and get a sequence of transformations and a sequence of systems of differential equations. For the latter, which we write as

$$\dot{\zeta}_{(k)} = X_{(k)}(\zeta_{(k)}) ,$$

we obtain

$$|\hat{X}_{(k)}|_{r_k, s_k} < s_k^{1+\alpha}$$

hence, by Cauchy's inequality

$$|\hat{X}_{(k)\eta}(\xi, 0)| < s_k^a$$

and it becomes clear that $X_{(k)}$ and its $\eta$ derivatives at $\eta = 0$ converge to zero as $k \to \infty$, leading to (1.13). We observe that the $\eta$ range shrinks to zero as $k \to \infty$ and the above estimates for $\lim_{k\to\infty} X_{(k)}$ hold only for $\eta = 0$.

To complete the proof of Theorem 5.1 we have to show that the composition of the infinitely many coordinate transformations converges. It will yield the desired transformation of Theorem 5.1. The convergence of this composed transformation follows easily from (3.17) with the choice $r = r_k$, $\rho = r_{k+1}$, $s = s_k$. Since $s_k$ tends to zero very rapidly while $r_k - r_{k+1}$ approaches zero only geometrically the right-hand side diminishes also rapidly and the composition of such mappings converges with its $\eta$ derivatives at $\eta = 0$ in $|\mathrm{Im}\, \xi_k| < \frac{r_0}{2}$. Since the mapping is linear in $\eta$ we obtain, in fact, convergence in $|\eta_k| < \frac{s_0}{2}$, say, and find the estimates

$$|\hat{w}|_{\frac{r_0}{2}, \frac{s_0}{2}} < c_9 r_0^{-\lambda} s_0^{1+a} = (c_9 r_0^{-\lambda} s_0^a) s_0 \ .$$

The expression in the parenthesis, can, like in (3.20), be made arbitrarily small by diminishing $\delta$ and since $s_0 = \delta$, we obtain the inequality (1.14).

d) *Proof of Lemma* 5.4

To estimate the error term $\hat{X}$ in the transformed equation we rewrite the defining equations (3.2), (3.3) of the transformed equation in more compact form. Using the notations from (3.7), (3.8), (3.9) we can combine (3.2), (3.3) to

$$\partial \hat{w} - \begin{pmatrix} \hat{v} \\ 0 \end{pmatrix} = \hat{w}_\xi(\xi, 0)\eta + (\hat{h})_1 \ . \tag{3.22}$$

On the other hand the transformation formula (1.16) takes the form

$$w_\zeta X = h(w) \ ,$$

or with

$$\chi = \begin{pmatrix} \omega + \eta \\ 0 \end{pmatrix} + \hat{\chi} \;, \qquad w = \zeta + \hat{w} \;, \qquad h = \begin{pmatrix} \omega + y \\ 0 \end{pmatrix} + \hat{h}$$

this takes the form

$$w_\zeta \hat{\chi} + \partial \hat{w} - \begin{pmatrix} \hat{v} \\ 0 \end{pmatrix} = -\hat{w}_\xi \eta + \hat{h}(w) \; .$$

Subtracting (3.22) from this we find

$$w_\zeta \hat{\chi} = -(\hat{w}_\xi - \hat{w}_\xi(\xi, 0))\eta + \hat{h}(w) - (\hat{h})_1 \; . \tag{3.23}$$

These equations will be used for an estimate of $\hat{\chi}$.

From the assumption (3.15) we obtain (3.17) from inequality (3.10) and, by Cauchy's estimate,

$$|\hat{w}_\zeta|_{\rho,\sigma} < c_5(r-\rho)^{-\lambda} s^a$$

since by (3.13) $r-\rho > s^{1-a} > s > s-\sigma$. By (3.14) the right-hand side of the last inequality is $< \frac{1}{2}$ and hence $w_\xi = I + \hat{w}_\zeta$ has an inverse, with norm $< 2$. Thus we conclude from (3.23)

$$|\hat{\chi}|_{\rho,\sigma} < 2|(\hat{w}_\xi - \hat{w}_\xi(\xi, 0))\eta|_{\rho,\sigma} + 2|\hat{h}(w) - (\hat{h})_1(w)|_{\rho,\sigma} + 2|(\hat{h})_1(w) - (\hat{h})_1(\zeta)|_{\rho,\sigma} \; . \tag{3.24}$$

We will estimate the 3 terms separately. Using the Cauchy estimate we find

$$|(\hat{w}_\xi - \hat{w}_\xi(\xi, 0))\eta|_{\rho,\sigma} \le |\hat{w}_{\xi\eta}|_{\rho,\sigma}\, \sigma^2 \le c_{10}(r-\rho)^{-\lambda-1} s^{2+a} \; . \tag{3.25}$$

If $0 < \sigma' < s$ we have

$$|\hat{h} - (\hat{h})_1|_{\rho,\sigma'} \le \frac{1}{2} |\hat{h}_{yy}|_{r,\sigma'}\, \sigma'^2 < s^{1+a} \left(\frac{\sigma'}{s-\sigma'}\right)^2$$

and since $v(\xi, \eta) = \eta + \hat{v}$ lies in

$$|y| < \sigma + c_5(r-\rho)^{-\lambda} s^{1+a} \quad \text{for} \quad |\mathrm{Im}\, \xi_k| < \rho \;, \quad |\eta_k| < \sigma$$

we take for $\sigma'$ the right-hand side of the last relation. Then we get

$$|\hat{h}(w) - (\hat{h})_1(w)|_{\rho,\sigma}$$
$$\leq c_{11}\, s^{1+\alpha}\left(\frac{\sigma}{s} + c_5(r-\rho)^{-\lambda}s^{\alpha}\right)^2 \leq c_{11}\, s^{1+\alpha}\left\{2\left(\frac{\sigma}{s}\right)^2 + 2[c_5(r-\rho)^{-\lambda}s^{\alpha}]^2\right\}$$

and since, by (3.14), the square bracket is $< 1$ we enlarge the right-hand side by omitting the square and get with (3.18)

$$|\hat{h}(w) - (\hat{h})_1(w)|_{\rho,\sigma} \leq c_{12}s^{1+\alpha}\left\{(r-\rho)^{-\lambda}s^{\alpha} + \left(\frac{\sigma}{s}\right)^2\right\} .$$

Finally,

$$|(\hat{h})_1(w) - (\hat{h})_1(\zeta)|_{\rho,\sigma} \leq \max|\hat{h}_z|\;|\hat{w}|_{\rho,\sigma} + \max|\hat{h}_{xy}|\;|\hat{w}|_{\rho,\sigma}|v|_{\rho,\sigma} .$$

In this relation we estimate $v$ with the aid of (3.13), (3.14) by

$$|v|_{\rho,\sigma} < \sigma + c_5(r-\sigma)^{-\lambda}s^{\alpha}s < \sigma + \frac{1}{2}\,s < s .$$

Furthermore, (3.13) implies $s < r-\rho$ which is used in the estimate

$$|(\hat{h})_1(w) - (\hat{h})_1(\zeta)|_{\rho,\sigma} \leq s^{\alpha}c_5(r-\rho)^{-\lambda}s^{1+\alpha} + \frac{s^{\alpha}}{r-\rho}\,c_5(r-\rho)^{-\lambda}s^{1+\alpha}s$$
$$\leq 2s^{\alpha}c_5(r-\rho)^{-\lambda}s^{1+\alpha} .$$

We observe that the term on the right-hand side is already present in the previous estimate. Moreover, the expression on the right-hand side of (3.25) is dominated by this term, since by (3.13) $s^{1-\alpha} < r-\rho$. Thus combining these estimates gives for (3.24)

$$|\hat{X}|_{\rho,\sigma} < c_6s^{1+\alpha}\left\{(r-\rho)^{-\lambda}s^{\alpha} + \left(\frac{\sigma}{s}\right)^2\right\}$$

with another constant $c_6$ and this is the assertion (3.16). This completes the proof of Lemma 5.4, and thus of Theorem 5.1.

## 4. *Generalizations*

### a) *Connection with Lie algebras*

The proof of Section 3 was carried out for reversible systems, however, as was pointed out, the methods extend to more general situations. For results about persistence of compact invariant manifolds, or subsystems on them, it is, generally, imperative that the class of differential equations is suitably restricted. The class of reversible or Hamiltonian vectorfields are examples. More generally, we denote by $\mathcal{K}$ a class of vectorfields defined, say in an open subset of Euclidean space. It is essential that this class $\mathcal{K}$ is preserved under a sufficiently wide group of coordinate transformations. If $\mathfrak{L}$ is the Lie algebra of this group we require that

$$(4.1) \qquad [\mathcal{K}, \mathfrak{L}] \subset \mathcal{K} ,$$

i.e., that the commutator $[K, L] = KL - LK$ belongs to $\mathcal{K}$ again, if $K \in \mathcal{K}$, $L \in \mathfrak{L}$. In our above examples $\mathcal{K}$ is represented by the vectorfields commuting with a reflection $\rho$ in the first case, canonical vectorfields in the second.

Without going into the details of the quantitative restrictions [99] we describe the formal algebraic conditions under which the method of the previous section is applicable and will illustrate it with a few examples.

We will single out an element $K_0$ in $\mathcal{K}$ which possesses an invariant submanifold. $K_0$ represents the unperturbed vectorfield, and we will have to impose some conditions on $K_0$ for the success of a perturbation theory. For this purpose we denote by $(\mathcal{K})_1$ the class of linearized differential operators, defined on the normal bundle of the invariant manifold. Furthermore, $\mathfrak{M}$ is assumed to be some linear subspace of $(\mathcal{K})_1$ with the following property: If $M$ is any element in $\mathfrak{M}$ sufficiently near zero then

$$(4.2) \qquad (\mathrm{ad}_{K_0+M}\mathfrak{L})_1 + \mathfrak{M} = (\mathcal{K})_1$$

where $\mathrm{ad}_X$ denotes the operator taking $L$ into $XL - LX$. If these conditions are amplified by some quantitative restrictions in appropriate norms it follows that for any $K$ sufficiently close to $K_0$ there exists a group element $g$ in the group $\mathcal{G} = \exp \mathfrak{L}$ generated by $\mathfrak{L}$, where $g$ is near the identity element, and an element $M_0 \in \mathfrak{M}$ such that

$$(g^*(K))_1 = K_0 + M_0 \tag{4.3}$$

where $g^*(K)$ denotes the vector field in $\mathcal{K}$ obtained from $K$ by transformation with $g$. The consideration of linearized vectorfields $(K)_1 \in (\mathcal{K})_1$ is forced upon us by the necessity for rapid convergence of the approximation which, in turn, requires that we control the linearized vectorfields during the iteration process.

Without trying to make these ideas rigorous we illustrate them with a few examples. First, it is evident that they are closely linked with concepts of Lie algebra. If $\mathcal{K} = \mathfrak{L}$ is a finite dimensional Lie algebra then $\mathfrak{M}$ plays the role of a Cartan subalgebra and $K_0$ that of a regular element. The case $\mathcal{K} \neq \mathfrak{L}$ is best illustrated with the situations where

$\mathcal{K}$ is represented by the Hermitean $n$ by $n$ matrices,

$\mathfrak{L}$ by the Lie algebra of anti-Hermitean $n$ by $n$ matrices,

$\mathfrak{M}$ by diagonal matrices

and $K_0$ is given by a diagonal matrix with distinct eigenvalues. Then ignoring the process of linearization $(\ )_1$ the conditions (4.1), (4.2) are readily verified and (4.3) corresponds to the fact, that any Hermitean matrix close to $K_0$ can be transformed by a unitary matrix $U$ into a diagonal matrix near $K_0$. This elementary fact does not require difficult tools, but our problems are concerned with the Lie algebras of infinite dimensions.

To cast the Theorem 2.9, proven in Section 3, into this framework we let

$$\mathcal{K} = \left\{ K = \sum_{k=1}^{n} f_x \partial_{x_k} + g_k \partial_{y_k};\ f(-x,y) = f(x,y),\ g(-x,y) = -g(x,y) \right\}$$

$$\mathfrak{L} = \left\{ L = \sum_{k=1}^{n} u_k \partial_{x_k} + v_k \partial_{y_k};\ u(-x,y) = -u(x,y),\ v(-x,y) = v(x,y) \right\}$$

$$\mathfrak{M} = (0),$$

and

$$K_0 = \sum_{k=1}^{n} (\omega_k + y_k)\partial_{x_k} ,$$

where $\omega_1, \omega_2, \ldots, \omega_n$ satisfy condition (1.10) again. The elements $K$ of $(\mathcal{K})_1$ have the same form as those of $\mathcal{K}$ but have coefficients $f_k$, $g_k$ linear in $y$. Moreover, the coefficients $f$, $g$, $u$, $v$ are assumed to be real analytic and of period $2\pi$ in the $x_k$. Then the conditions (4.1) and (4.2) hold; in fact, (4.2) corresponds to the solvability of the systems (3.2), (3.3). The group $\mathcal{G}$ generated by $\mathcal{L}$ consists of coordinate transformations commuting with the reflection $(x,y) \to (-x,y)$ and Theorem 2.9 is essentially given by (4.3). This problem is fairly simple since we have $\mathfrak{M} = (0)$.

One can modify the situation by observing that the frequencies themselves, $\omega_k$, do not have to be kept fixed, and it suffices to keep their ratios fixed. That leads to a result like that of Theorem 2.9 with condition (1.8) replaced by

$$\det\begin{pmatrix} F_y & \omega \\ \omega^T & 0 \end{pmatrix} \neq 0 . \tag{4.4}$$

Of course, then the frequencies of the resulting solutions will only be proportional to $\omega$. For the proof of this result one reduces the problem to $F(y) = \omega + Ay$ and would take $\mathcal{K}$, $\mathcal{L}$ as above but allow for $\mathfrak{M}$ vector-fields of the type

$$M = \sum_{k=1}^{n} (\lambda\omega_k + m_{k\ell} y_\ell)\partial_{x_k}$$

where $\lambda$, $m_{k\ell}$ are constants. The comparison of the constant terms in (4.2) leads to the solvability of an equation

$$Ac + \omega\lambda = b$$

for an n-vector $c$ and a scalar $\lambda$. Requiring $(c,\omega) = 0$ leads to the condition (4.4).

Finally, for the proof of Theorem 2.8 one can use the same setup. In that case $\mathcal{K}$ consists of Hamiltonian vectorfields

$$K = \sum_{k=1}^{n} \left(H_{y_k} \partial_{x_k} - H_{x_k} \partial_{y_k}\right)$$

with a Hamiltonian periodic in $x_1, \ldots, x_n$, while we take for Hamiltonian vectorfields

$$L = \sum_{k=1}^{n} \left(W_{y_k} \partial_{x_k} - W_{x_k} \partial_{y_k}\right)$$

where only $W - \sum_{k=1}^{n} c_k x_k$ is periodic with appropriate constants $c_1, c_2, \ldots, c_n$. The elements of $(\mathcal{K})_1$ are those $K \in \mathcal{K}$ for which $H$ contains $y$ at most quadratically. Furthermore, we take

$$K_0 = \sum_{k=1}^{n} \left(\omega_k + \sum_{\ell=1}^{n} a_{k\ell} y_\ell\right) \partial_{x_k}$$

where $A = (a_{k\ell})$ is a symmetric matrix and $\mathfrak{M}$ is given by

$$M = \sum_{k,\ell=1} m_{k\ell} y_\ell \partial_{x_k}, \quad m_{k\ell} = m_{\ell k}. \tag{4.5}$$

Thus, in the Hamiltonian case the matrix $A$ cannot be kept fixed under perturbation, if one works with canonical coordinate transformations. But it is to be observed that $M$ vanishes on $y = 0$ and therefore $M$ does not affect the invariance of the manifold $y = 0$ under $K_0$ or $K_0+M$.

b) *Invariant tori of lower dimensions*

So far we considered primarily invariant tori of a dimension half that of the phase space. The question arises whether one can extend such a perturbation theory to other dimensions. We restrict ourselves to Hamiltonian systems and discuss a result by S. Graff [100], according to which there is such a perturbation theory for invariant tori of dimension $2 \leq r \leq n$, if $2n$ is the dimension of the phase space. However, if $r < n$ the theory applies only to unstable tori.

To formulate the result we assume that the 2n-dimensional phase is described by the variables

$$x = (x_1, \dots, x_r) \ , \quad y = (y_1, \dots, y_r) \ , \quad p = (p_1, \dots, p_s) \ , \quad q = (q_1, \dots, q_s)$$

where $r+s = n$ and the invariant two-form

$$\sum_{\rho=1}^{r} dx_\rho \wedge dy_\rho + \sum_{\sigma=1}^{s} dp_\sigma \wedge dq_\sigma \ . \tag{4.6}$$

We assume that the Hamiltonian

$$H = H_0 + \mu H_1 + \dots \tag{4.7}$$

is real analytic in all variables and of period $2\pi$ in $x_\rho (\rho = 1, \dots, r)$ and that $H_0$ is of the form

$$H_0 = F(y) + G(y, p, q)$$

where

$$G(y, 0, 0) = 0 \ , \quad G_p(y, 0, 0) = 0 \ , \quad G_q(y, 0, 0) = 0 \ .$$

Then, for $\mu = 0$, we have invariant tori in

$$y = c \ , \quad p = 0 = q$$

on which the flow is given by

$$\dot{x} = F_y(c) \ .$$

The problem is to study the continuation of such a torus for small $\mu$.

THEOREM 5.2. *If, in the above notation,*

$$\det (F_{yy}) \neq 0 \quad \text{for} \quad p = q = 0$$

*and the matrix*

$$\begin{pmatrix} G_{qp} & G_{qq} \\ -G_{pp} & -G_{pq} \end{pmatrix}$$

*for* $p = q = 0$ *and* $y$ *in some open set has no eigenvalue on the imaginary axis then the system, defined by* (4.6), (4.7), *possesses* $r$*-dimensional tori as invariant submanifolds. In fact, the flow* $\dot{x}_\rho = \omega_\rho (\rho = 1, \ldots, r)$ *with frequencies satisfying a condition of type* (1.10) *can be analytically embedded as subsystems.*

This result can be applied to the equilibrium problem of a Hamiltonian system, when $2r$ pairs of eigenvalues $\pm\alpha_1, \ldots, \pm\alpha_r$ are purely imaginary and the remaining $2n-2r$ are not on the imaginary axis. Assuming

$$\sum_{\rho=1}^{r} j_\rho \alpha_\rho \neq 0 \quad \text{if} \quad 1 \leq |j| \leq 4$$

and imposing a nondegeneracy condition on the nonlinear terms one finds quasi-periodic solutions with $r$ independent frequencies for this system near the equilibrium.

Since these invariant tori are unstable one may inquire for the manifolds $W^{\pm}$ of asymptotic solutions approaching this torus for $t \to \pm\infty$. It turns out that they are also real analytic manifolds of dimension $n$, and they are Lagrangean manifolds in the sense that (4.6) vanishes on their tangent space.

The proof of Theorem 5.2 can be given along the lines of Section 4a). One can reduce the unperturbed Hamiltonian to the form

$$F = (\omega, y) + \frac{1}{2}(Ay, y) + (Bp, q)$$

where $A$, $B$ are constant matrices, $A$ an $r$ by $r$ matrix, $B$ an $s$ by $s$ matrix, and according to our assumptions, $A$ nonsingular, and $B$ with its eigenvalues in the right half plane.

For $\mathcal{K}$, $\mathcal{L}$ we take again Hamiltonian vectorfields of the type discussed before (§4a)) and for

$$K_0 = \sum_{\rho=1}^{r} F_{y_\rho} \partial_{x_\rho} + \sum_{\sigma=1}^{s} \left( F_{q_\sigma} \partial_{p_\sigma} - F_{p_\sigma} \partial_{q_\sigma} \right)$$

while $\mathfrak{M}$ consists of vectorfields

$$M = \sum_{\rho=1}^{r} \left(\Phi_{y_\rho} \partial_{x_\rho} - \Phi_{x_\rho} \partial_{y_\rho}\right) + \sum_{\sigma=1}^{s} \left(\Phi_{q_\sigma} \partial_{p_\sigma} - \Phi_{p_\sigma} \partial_{q_\sigma}\right)$$

with Hamiltonians of the form

$$\Phi = \frac{1}{2} (\hat{A}y, y) + (\hat{B}(x)p, q)$$

where $\hat{B}(x)$ is matrix depending periodically on $x$, while $\hat{A}$ is a constant matrix.

The new feature is that $K_0 + M$ does not have constant coefficients anymore and the verification of (4.2) requires the solution of first order partial differential equations with variable coefficients, which is possible if the eigenvalues of $B$ are in the right half plane.

c) *Differentiable case*

The above technique has been described only for the case of real analytic functions. It can be generalized to sufficiently often differentiable functions but gets more complicated in that case. That one needs a number of derivatives to start with is clear by looking at the operator (2.4): One can show, under condition (1.10), $\Lambda$ maps $C^\ell$ continuously into $C^m$ if $\ell > \tau + m$. Here $C^\ell$ for $\ell = [\ell] + \alpha$ $(0 \leq \alpha < 1)$ stands for functions whose derivatives of order $[\ell]$ are Hölder continuous with exponent $\alpha$. Of course, solving equations of the type (3.2), (3.3) gives rise to a larger, but finite derivative loss.

The difficulty arises if one applies the iteration and, without other precaution, one would lose all differentiability after finitely many steps. This is avoided by approximating the functions by smoother ones and keeping careful control of the error. An example of this approximation was used in Section 2 when we replaced $\hat{f}(\xi)$ by $\hat{f}_1(\xi)$ (see (2.9)) since that method actually also contains a derivative loss. For the proof of the differentiable analogue of Theorem 2.9, for example, one needs more refined approximation techniques whose basic characteristic is that smoother functions are approximated more closely. An example is truncation of Fourier series. In the analytic case the role of the truncation is played by the shrinking of the complex domain.

We do not want to describe the method in the differentiable case here and refer to [63], [66]. We just point out that the existence theorems require a certain minimal number of differentiability, see the result of Takens [67] quoted in Chapter II. Rüssman succeeded in carrying out proofs of this type with very mild differentiability requirements [65]. In [60] it was shown that Theorem 2.8 can be established in the class $C^{\ell}$ with $\ell > 2r + 2$ of Hamiltonian. This result was achieved by admitting for elements of the form (4.5) where, however, $m_{k\ell}$ are periodic functions of $x$. This means effectively that we do not insist on constant coefficients for the linearized vectorfield. While this makes it harder to solve the equations represented by (4.2) it also leads to milder differentiability assumptions.

d) *A third order partial differential equation*

We mentioned that the above method is based largely on transformation theory. To illustrate that this is essential for overcoming the small divisor difficulty we contrast the two partial differential equations

$$\omega_1 u_{x_1} + \omega_2 u_{x_2} + \mu a(x)u = g(x) \text{ (mod const.)} \tag{4.8}$$

with

$$\omega_1 u_{x_1x_1x_1} + \omega_2 u_{x_2x_2x_2} + \mu a(x)u = g(x) \text{ (mod const.)} \tag{4.9}$$

where all functions are assumed to be real analytic of period $2\pi$ in $x_1$, $x_2$ and

$$a(-x) = -a(x) \ . \tag{4.10}$$

On the right-hand side arbitrary additive constants are allowed, which for $\mu = 0$ correspond to the vanishing of the mean-value of the right-hand side. The ratio $\omega_1/\omega_2$ is assumed to be irrational and a condition (1.10) is required. Thus for $\mu = 0$ the above equations are solvable according to our previous considerations, since by (1.10)

$$|j_1^3\omega_1 + j_2^3\omega_2| \geq \gamma(|j_1^3| + |j_2|^3)^{-\tau} \geq \gamma|j|^{-\tau} .$$

But, while (4.8) admits a solution for all sufficiently small $\mu$ our methods fail for (4.9) as we will indicate. The difference is that (4.8) as a first order partial differential operator admits a transformation theory which is not the case for the third order equation (4.9).

In both cases we have a small divisor problem and the eigenvalues of the linear operators on the left cluster at 0 and it is a problem that 0 does not become an eigenvalue for small values of $\mu$. For (4.8) this is the case since the spectrum is independent of $\mu$. Setting $\partial = \omega_1\partial_{x_1} + \omega_2\partial_{x_2}$ we solve

$$\partial b = a(x)$$

where $b(x)$ is an even function of mean value 0. Then we have for any real analytic periodic function $v(x)$

$$(\partial + \mu a)(e^{-\mu b}v) = e^{-\mu b}\partial v$$

which reduces the problem to inverting $\partial$. It also shows that $v \to e^{-\mu b}v$ transforms $\partial + \mu a$ into

$$e^{\mu b}(\partial + \mu a)\, e^{-\mu b} = \partial$$

so that the spectrum is independent of $\mu$, even without smallness conditions on $\mu$.

Thus (4.8) can be solved easily. Actually, the same is true for

$$(I\partial + \mu A(x))u = g(x) \text{ (mod const.)}$$

where $u$, $g$ are m-vectors, $A(x) = -A(-x)$ and $I$ are $m$ by $m$ matrices, $I$ is the identity matrix. In this case the above exponential trick is not applicable unless the $A(x)$ commute for different $x$ values, but an iteration of transformations reducing the size of $A(x)$ does succeed also in this case. Again there exists an invertible matrix $U(x)$ satisfying

$$(\partial+\mu A)U = 0$$

such that

$$U^{-1}(\partial+\mu A)U = \partial \ .$$

On the other hand this approach is not applicable to (4.9) and one may conjecture that (4.9) cannot be solved for all sufficiently small values of $\mu$. Indeed the spectrum of the operator on the left of (4.9) has a point spectrum dense on the imaginary axis and, if these eigenvalues vary with $\mu$ one has to expect that for sufficiently small $\mu$ some eigenvalue will pass through zero. Actually we do not know the nature of the spectrum of this operator for $\mu \neq 0$ but if one proceeds formally with perturbation theory, say for

$$a(x) = \sin(m,x) = \sin(x_1+x_2) \ , \qquad m = (1,1)$$

one finds for the eigenvalues

$$\lambda(j) = i(j_1^3\omega_1 + j_2^3\omega_2)$$

a perturbation of the form

$$\lambda = \lambda(j) + \mu^2 q + 0(\mu^3)$$

where

$$q = -\frac{3}{2}(j,\omega)(\lambda(j+m) - \lambda(j))^{-1}(\lambda(j-m) - \lambda(j))^{-1} \ .$$

Since $q$ does not vanish, in general, one may expect the spectrum to change under perturbation and thus destroy the solvability of (4.9). However, this argument is not conclusive since a perturbation theory for (4.9) is not available and it is conceivable that the spectrum is continuous.

In any event (4.9) represents an equation not accessible to our method. The main difference is that the spectrum of (4.8), even for $\mu \neq 0$, is an additive group generated by only two generators, while a third order operator does not have such a rigid structure for its spectrum.

## A. APPENDIX TO CHAPTER V

a) *Rate of convergence for the scheme of* §2b)

In this section we wish to investigate the rate of convergence for the method for finding a root of a function discussed in §2c). We will merely determine the exponent of convergence and not make the constants explicit.

We go back to formula (2.15) which defines two sequences $x_k$, $a_k$ where $x_k \to \alpha$, $a_k \to f'^{-1}(\alpha)$ where $f(\alpha) = 0$ under appropriate smallness conditions. We introduce the quantities

$$y_k = x_k - \alpha \ , \qquad z_k = 1 - f'(x_k)\, a_k$$

which tend to zero. If we use the Taylor expansion of $f$ at $x_k$ we find

$$0 = f(\alpha) = f(x_k) - f'(x_k)\, y_k + O(y_k^2)$$

so that with the first equation of (2.15) we find

$$\begin{aligned} y_{k+1} &= y_k - a_k(f'(x_k)\, y_k + O(y_k^2)) \\ &= (1 - a_k f'(x_k))\, y_k + O(y_k^2) \\ &= z_k y_k + O(y_k^2) \end{aligned}$$

or, for $|y_k| + |z_k|$ sufficiently small,

(A.1) $$|y_{k+1}| \leq |z_k y_k| + c_1 |y_k|^2 \ .$$

To estimate $z_{k+1}$ we write the second equation of (2.15) in the form

$$1 - a_{k+1} f'(x_k) = (1 - a_k f'(x_k))^2 = z_k^2 \ .$$

Thus

$$z_{k+1} = -a_{k+1}(f'(x_{k+1}) - f'(x_k)) + z_k^2$$

$$= -a_{k+1}f''(y_{k+1} - y_k) + z_k^2$$

where $f''$ is to be evaluated at an intermediate point. If we ignore terms of order $y_k^2 + z_k^2$ we can, by (A.1), drop $y_{k+1}$. Since

$$a_{k+1} = a_k(2 - f'(x_k)\,a_k)$$

$$= f'^{-1}(\alpha) + O(|y_k| + |z_k|)$$

we find

$$z_{k+1} = Ay_k + O\left(y_k^2 + z_k^2\right), \quad A = \frac{f''(\alpha^*)}{f'(\alpha)}$$

with an intermediate point $\alpha^*$; or

$$\text{(A.2)} \qquad |z_{k+1}| \le |A|\,|y_k| + c_1\left(y_k^2 + z_k^2\right).$$

To study the convergence rate of (A.1), (A.2) we introduce a sequence $\varepsilon_0, \varepsilon_1, \varepsilon_2, \ldots$ of positive, monotone decreasing numbers satisfying

$$\text{(A.3)} \qquad \begin{cases} \varepsilon_{k+2} = c_2\varepsilon_k\varepsilon_{k+1} & \text{for} \quad k = 0, 1, 2, \ldots \\ \varepsilon_1 \ge c_2\varepsilon_0^2 \end{cases}$$

with

$$c_2 = \max\{c_1(1 + 4A^2)|A|^{-1};\ c_1 + 2|A|\}.$$

Then we claim that the inequalities

$$\text{(A.4)} \qquad |y_k| < \varepsilon_{k+1}, \quad |z_k| < 2|A|\,\varepsilon_k$$

hold for all $k \ge 1$ if they hold for $k = 0$. By (A.3) we obtain for $k \ge 1$

$$\varepsilon_k^2 \le \varepsilon_k\varepsilon_{k-1} = c_2^{-1}\varepsilon_{k+1},$$

a relation which by (A.3) holds for $k = 0$ also. If, now, (A.4) holds for some $k \geq 0$ we conclude from (A.2)

$$|z_{k+1}| \leq |A|\varepsilon_{k+1} + c_1(1+4A^2)\varepsilon_k^2$$
$$\leq \left(|A| + \frac{c_1}{c_2}(1+4A^2)\right)\varepsilon_{k+1} \leq 2|A|\varepsilon_{k+1} \, .$$

Here we used the definition of $c_2$. Thus the second relation (A.4) is verified for $k+1$. Similarly, (A.1) implies

$$|y_{k+1}| \leq 2|A|\varepsilon_k\varepsilon_{k+1} + c_1\varepsilon_{k+1}^2$$
$$\leq (2|A| + c_1)\varepsilon_k\varepsilon_{k+1} \leq c_2\varepsilon_k\varepsilon_{k+1} = \varepsilon_{k+2}$$

which verifies the first relation (A.4) for $k+1$ in place of $k$.

It is easy to study the behavior of the sequence $\varepsilon_k$. In fact, one sees that

$$\varepsilon_k = c_2^{-1}\,\vartheta^{\alpha_k} \; ; \quad 0 < \vartheta < 1$$

is a solution if

$$\alpha_{k+2} = \alpha_{k+1} + \alpha_k$$

where we choose $\alpha_0 = \alpha_1 = 1$. These numbers behave asymptotically like

$$\alpha_k \sim c_3\left(\frac{\sqrt{5}+1}{2}\right)^k$$

which establishes $\kappa = \frac{1}{2}(\sqrt{5}+1)$ as the exponent of convergence.

b) *The improved scheme by Hald*

If we replace the recursion formulae (2.15) by

$$x_{k+1} = x_k - a_k f(x_k)$$
$$a_{k+1} = a_k + a_k(1 - f'(x_{k+1})a_k)$$

where just the argument in $f'$ was changed, then we have quadratic convergence. This observation is due to O. H. Hald.

To prove this statement we use the same notation as above and note that (A.1) remains true without change. But the estimate for $|z_{k+1}|$ gets improved since now

$$z_{k+1} = (1 - a_{k+1} f'(x_{k+1})) = (1 - a_k f'(x_{k+1}))^2 .$$

We estimate

$$\begin{aligned} 1 - a_k f'(x_{k+1}) &= z_k - a_k(f'(x_{k+1}) - f'(x_k)) \\ &= z_k - a_k f''(y_{k+1} - y_k) \\ &= O(|y_k| + |z_k|) . \end{aligned}$$

Hence, (A.2) is replaced by

$$|z_{k+1}| \le c_4\left(y_k^2 + z_k^2\right) .$$

Combining (A.1) with this relation we have for $\delta_k = \sqrt{y_k^2 + z_k^2}$

$$\delta_{k+1} \le c_5 \delta_k^2 \quad \text{for} \quad k = 0, 1, \ldots ,$$

which makes quadratic convergence evident.

# CHAPTER VI

# PROOFS AND DETAILS FOR CHAPTER III

## 1. *Outline*

In the following 6 sections we provide the details for the proof of Theorems 3.5 and 3.6 of Chapter III. As was stated there already this model was treated by Sitnikov, but the results described here are mainly those of Alekseev. However, the method of proof given here is new. The following exposition is based on a device suggested by R. McGehee according to which one finds a hyperbolic periodic orbit at infinity. From this orbit issue two invariant manifolds which intersect on some homoclinic point. Thus the situation is essentially the same as in the case of a homoclinic point. Actually there are technical differences, as the periodic orbit at infinity is only hyperbolic in a degenerate sense. Also the choice of the section which determines the mapping will be different from the one near a homoclinic point. Nevertheless, the idea of a periodic orbit at infinity aids the geometrical intuition, and the coordinates introduced by McGehee simplifies the calculation.

We want to point out some details: The smallness of the eccentricity $\varepsilon$ is not essential in the whole proof except for Lemma 3, for the verification that the angle of intersection of the boundaries is not zero. But since this angle is a real analytic function of $\varepsilon$ in $0 < \varepsilon < 1$ it can vanish only for a discrete set of $\varepsilon$ values. If one excludes these all results hold for $0 < \varepsilon < 1$.

In §7 we give a discussion of Theorem 3.7 about embedding the shift near a homoclinic point. In most treatments on this subject a theorem of Hartman is used; this asserts that a mapping near a hyperbolic fixed point is a linear map in appropriate coordinates. However, in general, the transformation between the coordinate systems is merely a homeomorphism, and can be assumed to be $C^1$ only under additional conditions on the eigenvalues. See Hartman [51], p. 245-246. In fact, the use of such a linearizing transformation is quite unnecessary, as out treatment in §7 shows. Finally, in §8 we discuss the nonexistence of real analytic integrals near a homoclinic point, showing, in particular, that the restricted three-body problem does not admit such an integral.

2. *Behavior near infinity*

a) *Escaping solutions*

The differential equations to be studied are

$$\ddot{z} + \frac{z}{(z^2 + r^2)^{\frac{3}{2}}} = 0 \tag{2.1}$$

where

$$r = \frac{1}{2}(1 - \varepsilon \cos t) + O(\varepsilon^2) \tag{2.2}$$

is a real analytic function of $\varepsilon$, $t$ satisfying

$$r(t + 2\pi) = r(t) = r(-t) . \tag{2.3}$$

We consider solutions of the above differential equation with initial values

$$z(t_0) = 0 , \quad \dot{z}(t_0) > 0 \tag{2.4}$$

and denote by $t_1$ the smallest number $> t_0$ for which $z(t_1) = 0$ if it exists; otherwise we set $t_1 = \infty$. We visualize the initial values $(\dot{z}(t_0), t_0)$ as polar coordinates of a plane $R^2$, $t_0$ representing the angle and $\dot{z}(t_0) \geq 0$ the radius. In accordance with Chapter III we denoted with $D_0$ the set of points in the plane for which $t_1 < \infty$ to which we add the point $(0, t_0)$. Thus $D_0$ represents the initial values of the orbits which return to $z = 0$. Obviously $D_0$ is an open set.

We turn to the study of the complement of $D_0$. Thus let $z(t)$ be a solution of (2.1) satisfying (2.4) and $t_1 = \infty$, hence $\dot{z}(t) > 0$ for $t > t_0$. Thus $z$ is monotonically increasing for $t > t_0$ and, from (2.1), we conclude that

$$z(t) \to \infty \quad \text{as} \quad t \to \infty .$$

Since, again by (2.1), $\ddot{z} < 0$ for $t > t_0$ the function $\dot{z} > 0$ is monotonically decreasing and

$$\dot{z}(\infty) = \lim_{t \to \infty} \dot{z}(t) \geq 0$$

exists. Usually one refers to an orbit as a hyperbolic orbit if $\dot{z}(\infty) > 0$ and as a parabolic one if $\dot{z}(\infty) = 0$. But to avoid confusion with the concept of hyperbolic fixed points or hyperbolic periodic orbits we avoid this notation. Our aim will be to describe the asymptotic behavior of $z(t)$ in the two cases $\dot{z}(\infty) = 0$ and $\dot{z}(\infty) > 0$.

b) *McGehee's transformation*

For this purpose we make the coordinate transformation, due to McGehee

$$z = \frac{2}{q^2}, \quad \dot{z} = -p, \quad dt = 4q^{-3}ds \tag{2.5}$$

for $0 < q < \infty$ so that $q \to 0$ corresponds to $z \to \infty$. The differential equation (2.1) takes the form

$$\frac{dq}{ds} = p, \quad \frac{dp}{ds} = q\left(1 + \frac{q^4}{4} r^2\right)^{-\frac{3}{2}}, \quad \frac{dt}{ds} = \frac{4}{q^3} . \tag{2.6}$$

If we would ignore the t-dependence of $r$ we could consider $p = q = 0$ as a hyperbolic point for this differential equation, for which the stable and unstable manifolds are given by $q = \pm p + \ldots$, two power series in $p$. Since, however, $r$ is a periodic function of $t$ we may consider $p = q = 0$ as a periodic orbit through which there pass two invariant manifolds

$$q = X(p, t) = p(1 + a_4 p^4 + a_7(t) p^7 + \ldots)$$

and

$$q = X(-p, -t)$$

where $X$ has period $2\pi$ in $t$ and is real analytic for $0 < p < a$ for some positive $a$. At $p = 0$ one can assert only that $X$ is in $C^\infty$.

The existence proofs for these manifolds can be reduced to the existence of invariant curves for the corresponding mapping, following the orbits from $t = 0$ to $t = 2\pi$. This is a degenerate fixed point since both eigenvalues of the Jacobian are 1 but the existence proofs for the invariant curves can be extended to this case as was shown in [101], [102]. Here we just point out that the above expansion of $X$ can be computed by comparison of coefficients, after the existence has been established, and one finds

$$a_4 = \frac{1}{32\pi} \int_0^{2\pi} r^2(t)dt$$

although this will not be used in the following. Also, since (2.5) takes the reflection $(z, \dot{z}, t) \to (z, -\dot{z}, -t)$ into

$$(q, p, t) \to (q, -p, -t) \tag{2.7}$$

it suffices to establish the existence of *one* invariant manifold.

To transform the invariant manifold in the coordinate planes we set

$$\begin{cases} x = \frac{1}{4}(q - X(-p, -t)) = \frac{1}{4}(q + p) + \ldots \\ y = \frac{1}{4}(q - X(p, t)) = \frac{1}{4}(q - p) + \ldots \end{cases}$$

and obtain for the differential equations

$$\begin{cases} \dfrac{dx}{ds} = x(1 + O_4) \\ \dfrac{dy}{ds} = -y(1 + O_4) \\ \dfrac{dt}{ds} = \dfrac{1}{2}\{(x + y)^3 + O_4\}^{-1} = 4q^{-3} . \end{cases} \tag{2.8}$$

Here $O_n$ stands for a $C^\infty$ function $f = f(x, y, t)$ of period $2\pi$ in $t$ and such that $\lambda^{-n} f(\lambda x, \lambda y, t)$ is bounded uniformly in $t$ for $\lambda \to 0$ where $\lambda$ runs through positive values. The above differential equation is restricted to the domain

$$q = 2(x+y) + O_4 > 0 \tag{2.9}$$

and $q = 0$ corresponds to $z = \infty$.

c) *Orbits near singularity*

To discuss the orbits near $x = y = 0$ we eliminate the variable $s$ and write

$$\begin{aligned} \dot{x} &= f(x, y, t) = x(2(x+y)^3 + O_4) \\ \dot{y} &= g(x, y, t) = y(-2(x+y)^3 + O_4) \ . \end{aligned} \tag{2.10}$$

Since the reflection (2.7) is transformed into $(x, y, t) \to (y, x, -t)$ we have

$$-g(x, y, t) = f(y, x, -t) \ .$$

LEMMA 1. *Let* $x(t)$, $y(t)$ *for* $t \geq t_0$ *be a solution in* (2.9) *and in a sufficiently small neighborhood* $|x|, |y| < a$. *Such a solution has* $x = y = 0$ *as a cluster point if and only if* $x(t_0) = 0$. *Moreover, all such solutions on* $x = 0$ *tend to the origin.*

*Proof.* For $x \neq 0$

$$\frac{f(x, y, t)}{x} = \frac{q^3}{4}(1 + O_4) \geq \frac{q^3}{8} > 0$$

in a sufficiently small neighborhood. Hence if $x(t_0) \neq 0$ we have

$$\frac{\dot{x}}{x} \geq \frac{1}{8} q^3 \tag{2.11}$$

as long as $x(t) \neq 0$. We conclude that

$$|x(t)| \geq |x(t_0)| > 0 \quad \text{for} \quad t > t_0$$

and such solutions cannot have the origin as cluster point.

On the other hand, if $x(t_0) = 0$ then $x(t) = 0$ for $t > t_0$ and the differential equation reduces to

$$\dot{y} = g(0, y, t) = y(-2y^3 + O_4) \leq -y^4$$

if a is sufficiently small. Integrating we have

$$y(t) \leq y_0\{1 + y_0^3(t - t_0)\}^{-\frac{1}{3}} \to 0 \quad \text{for} \quad t \to \infty$$

which proves the statement.

From this result it is easy to discuss the local behavior of the solution near $x = y = 0$ in the domain (2.9). From the Lemma 1 it follows that the only solutions approaching the origin lie on $x = 0$. If initially $x(t_0) > 0$ then $x(t) \geq x(t_0)$. We claim that all these solutions must leave our neighborhood. Indeed, if $y(t_0) = y_0 \geq 0$ then $y(t) \geq 0$ for $t > t_0$ and

$$q(x, y, t) = 2(x + y) + O_4 \geq 2x + O_4(x) \geq x \geq x_0$$

and from (2.11) we see that $\dot{x} \geq \frac{1}{8} x_0^4$ which shows that $x$ escapes our neighborhood. If, on the other hand, $y_0 < 0$ then $y(t) < 0$ for $t > t_0$. Hence from (2.6)

$$\frac{dq}{dt} = \frac{q^3}{4} p = \frac{q^3}{2}(x - y + \ldots) > 0$$

and $q > q_0$ for $t > t_0$. Thus by (2.11) $\dot{x} \geq \frac{1}{8} x q_0^3$ which shows that the solution escapes our neighborhood.

On the other hand, if $x(t_0) < 0$ then the solution approaches a definite point on $q = 0$ as $t \to \infty$. For this we observe that $x$ is monotonically decreasing; the same is shown for $y(t)$ so that both $x$, $y$ approach a limit, since $x$, $y$ both are bounded from below. It is clear from the differential equations that this limit point lies on $q = 0$.

Thus the orbits with $x(t_0) > 0$ correspond to orbits returning (elliptic type) those with $x(t_0) = 0$ escape with fixed velocity zero (parabolic type) and those with $x(t_0) < 0$ correspond to excaping orbits with positive

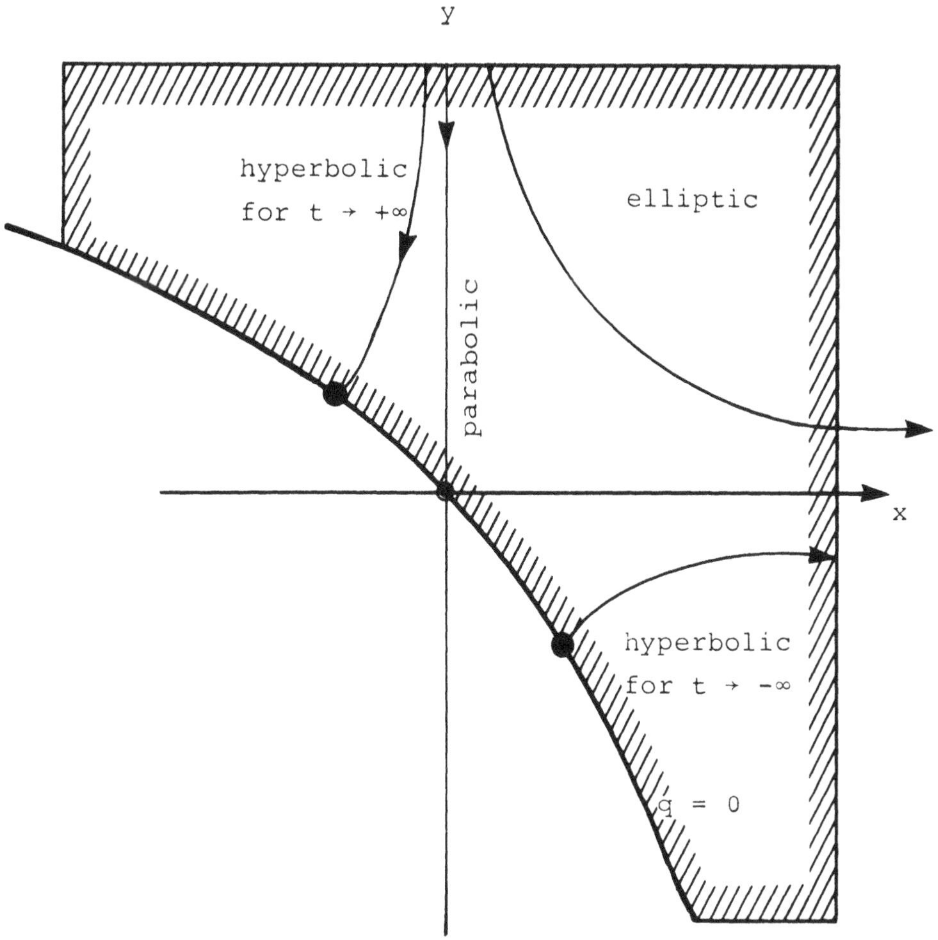


Fig. 19

velocity at infinity (hyperbolic type). See Fig. 19. One can use the above formulae to derive asymptotic description for the escaping solutions: If $x(t_0) = 0$ hence $x(t) = 0$ for $t > t_0$, then (2.10) reduces to

$$\dot{y} = -2y^4 + O(y^5)$$

and integration yields

$$y = (6t)^{-\frac{1}{3}} + O\left(t^{-\frac{2}{3}}\right) \quad \text{as} \quad t \to +\infty .$$

Going back to the variables q, p we get

$$q = 2(x+y) + \ldots = 2y + \ldots = 2(6t)^{-\frac{1}{3}}\left(1 + O\left(t^{-\frac{1}{3}}\right)\right)$$

$$p = 2(x-y) - \ldots = -2y - \ldots = -2(6t)^{-\frac{1}{3}}\left(1 + O\left(t^{-\frac{1}{3}}\right)\right)$$

and finally

$$\begin{cases} z = 2q^{-2} = \frac{1}{2}(6t)^{\frac{2}{3}}\left(1 + O\left(t^{-\frac{1}{3}}\right)\right) \\ \dot{z} = -p = 2(6t)^{-\frac{1}{3}}\left(1 + O\left(t^{-\frac{1}{3}}\right)\right). \end{cases}$$

For the hyperbolic case, one has $\dot{z}(t) \to \dot{z}(\infty) > 0$ and it is easily verified that

$$z(t) = \dot{z}(\infty)t + \text{const} + O(t^{-1})$$

$$\dot{z}(t) = \dot{z}(\infty) + O(t^{-1})$$

as $t \to \infty$.

3. *Proof of Lemmas* 1 *and* 2 *of Chapter* III

a) We use the notation of Chapter III, §5, c) and characterize the solutions by their initial values $t = t_0$, $v_0 = |\dot{z}(t_0)|$ for $z(t_0) = 0$, and consider $v_0$, $t_0$ as polar coordinates in the plane $z = 0$. The set $D_0$ of initial values $v_0$, $t_0$ for which there exists another zero $t_1 > t_0$ of $z(t)$ is certainly an open set. The complement consists of solutions $z(t)$ which excape monotonically for $t \to \infty$. We divided these solutions into those with $|\dot{z}(\infty)| = 0$ (parabolic) and with $|\dot{z}(\infty)| > 0$ (hyperbolic). It is easily seen that the latter correspond to an open set in the $t_0$, $v_0$-plane. Finally, we will show that the parabolic orbits correspond to a simple closed curve, which by Jordan's curve theorem decomposes the plane into two components: The interior, given by $D_0$, and the exterior corresponding to the hyperbolic orbits.

b) To show that the (parabolic) orbits escaping for $t \to +\infty$ with $\dot{z}(\infty) = 0$ intersect the z-plane in a simple closed curve we only have to observe that in the coordinates $x, y$ of the previous section these orbits are characterized by $x = 0$, $y > 0$ and $t$ arbitrary. In particular, the plane $y = a > 0$ is intersected transversally by those orbits on the curve $x = 0$, $t$ arbitrary. Since $t$ is an angular variable this represents a simple closed curve, real analytic in the variables $x, y, t$. We obtain the desired curve in the plane $z = 0$ by subjecting the above curve to the mapping which is obtained by following the solutions from the surface $y = a$ for decreasing time to the plane $z = 0$. Since the orbits intersect these surfaces transversally it is clear that this mapping $(x, t) \to (\dot{z}_0, t_0)$ is real analytic for $x$ near $0$. By the uniqueness theorem for differential equations it is a diffeomorphism and $(x, t + 2\pi)$ goes into $(\dot{z}_0, t_0 + 2\pi)$ if $(x, t) \to (\dot{z}_0, t_0)$. This shows that the family of orbits, parabolic for $t \to +\infty$, intersects $z = 0$ in a simple closed, real analytic curve $\partial D_0$, whose interior $D_0$ corresponds to initial values of orbits returning to $z = 0$ for increasing $t$. Thus $D_0$ is the domain of definition of the mapping $\phi$ of Chapter III, §5, c). The orbits with initial values outside $D_0$ correspond to escape orbits, those of $\partial D_0$ to parabolic, the other to hyperbolic ones. This proves Lemma 1 of Chapter III, §5, c).

c) To represent the mapping $\phi$ analytically let $(v_0, t_0)$ be the polar coordinates of a point in $D_0$ and

$$z = z(t;\, v_0, t_0)$$

the solution of (2.1) with initial values

$$z(t_0;\, v_0, t_0) = 0\ , \quad \dot{z}(t_0;\, v_0, t_0) = v_0\ .$$

Then the image point $(v_1, t_1)$ of $(v_0, t_0)$ under $\phi$ is obtained as follows: $t_1$ is the next zero $> t_0$ of $z(t;\, v_0, t_0)$ and

$$v_1 = |\dot{z}(t_1;\, v_0, t_0)|\ .$$

Similarly, $\phi^{-1}$ takes $(v_0, t_0)$ into $(v_{-1}, t_{-1})$ where $t_{-1}$ is the nearest zero $< t_0$ of $z(t; v_0, t_0)$, if it exists, and

$$v_{-1} = |\dot{z}(t_{-1}; v_0, t_0)| .$$

To determine the domain of definition of $\phi^{-1}$ we observe that the differential equation (2.1) is invariant under the reflection $(z, \dot{z}, t) \to (z, -\dot{z}, -t)$ as $r(-t) = r(t)$, and this implies the identity

$$z(-t; -v_0, -t_0) = z(t; v_0, t_0) .$$

Thus if $\rho$ denotes the reflection $(v_0, t_0) \to (v_0, -t_0)$ we conclude that

$$\phi^{-1} = \rho^{-1}\phi\rho$$

and $\phi^{-1}$ is defined precisely in $\rho(D_0)$. Clearly, this is also the image domain of $\phi$ and we set $D_1 = \phi(D_0)$, so that $D_1 = \rho(D_0)$. This proves Lemma 2, aside from the area-preserving character. For this we write the differential equation (2.1) in Hamiltonian form

$$\dot{z} = H_v , \quad \dot{v} = -H_z$$

with

$$H = \frac{1}{2} v^2 - \frac{1}{\sqrt{z^2 + r^2(t)}} .$$

It is well known that the differential form (of Cartan)

$$dv \wedge dz - dH \wedge dt$$

is preserved under the flow and our mapping $\phi$ preserves the restriction of the above form to $z = 0$, $dz = 0$ which is given by

$$-dH \wedge dt = -d \frac{v^2}{2} \wedge dt = -v dv \wedge dt$$

as was claimed.

4. *Proof of Lemma* 3 *of Chapter* III

a) The proof of Lemma 3 is just based on a calculation of the boundary curve $\partial D_0$ which we write as

$$v_0 = \lambda(t_0, \varepsilon)$$

where, according to Lemma 1 the function $\lambda$ is a real analytic function of $t_0$ of period $2\pi$, and $\varepsilon$ denotes the eccentricity in (2.2). The boundary $\partial D_1$ of the range is given by

$$v_0 = \lambda(-t_0, \varepsilon)$$

and so both curves intersect for $t_0 = 0$. In order that the tangents at this point are different we need that

$$\left.\frac{\partial\lambda}{\partial t_0}\right|_{t_0=0} \neq 0 \; . \tag{4.1}$$

We shall show that

$$\frac{\partial^2\lambda}{\partial t_0 \partial\varepsilon} < 0 \quad \text{for} \quad t_0 = 0 \; , \quad \varepsilon = 0 \tag{4.2}$$

which implies (4.1) for sufficiently small $\varepsilon > 0$. In fact, as $\lambda$ is a real analytic function $\varepsilon$ in $0 \le \varepsilon < 1$ it follows that (4.1) can be violated only for a discrete set of values.

b) With

$$U(z, t) = (z^2 + r^2(t))^{-\frac{1}{2}} \tag{4.3}$$

we write the differential equations in the form

$$\ddot{z} = \frac{\partial U}{\partial z} \; .$$

From this we obtain an integral equation for the parabolic solutions, characterized by $\dot{z}(\infty) = 0$, whose existence is already established. For this we multiply the last equation by $-\dot{z}$ and integrate from $t$ to $\infty$ to get

(4.4)
$$\frac{1}{2}\dot{z}^2 = -\int_t^{\infty} \frac{\partial U}{\partial z}\dot{z}dt .$$

The solution with $z(t_0) = 0$, $\dot{z}(t_0) > 0$ defines a parabolic solution, say $z = z(t;\, t_0, \varepsilon)$, and
$$v_0 = \lambda(t_0, \varepsilon) = \dot{z}(t_0;\, t_0, \varepsilon) .$$

We will determine only the constant and linear terms in $\varepsilon$ of $z$ and $\lambda$. Setting $\varepsilon = 0$ in (4.4) we get

(4.5)
$$\frac{1}{2}\dot{z}^2 = -\int_t^{\infty} \frac{\partial U_0}{\partial z}\dot{z}dt = U_0(z)$$

where with (2.2) and (4.3)

(4.6)
$$\begin{cases} U = U_0 + \varepsilon U_1 + O(\varepsilon^2) \\ U_0 = \left(z^2 + \frac{1}{4}\right)^{-\frac{1}{2}} ; \quad U_1 = \frac{1}{4}\dfrac{\cos t}{\left(z^2 + \frac{1}{4}\right)^{\frac{3}{2}}} . \end{cases}$$

The differential equation (4.5) is independent of $t$ and can be integrated explicitly. We do not need the explicit expression though and just introduce the unique solution $z = \zeta(t)$ of (4.5) satisfying $\zeta(0) = 0$, $\dot{\zeta}(0) > 0$. Then $\zeta(t)$ is a positive, monotone increasing function for $t > 0$ with $\dot{\zeta}(0) = 2$ and the general solution of (4.5) is given by $\zeta(t - t_0)$. Therefore we can write
$$z(t;\, t_0, \varepsilon) = \zeta(t - t_0) + \varepsilon z_1 + O(\varepsilon^2)$$
and

(4.7)
$$\lambda(t_0, \varepsilon) = \dot{z}(t_0;\, t_0, \varepsilon) = \dot{\zeta}(0) + \varepsilon \dot{z}_1\Big|_{t=t_0} + O(\varepsilon^2) .$$

Since $\dot{\zeta}(0) = 2$ the curve $\partial D_0$ is in first approximation a circle of radius 2, in agreement with the discussion in Chapter III, §5, d).

To compute $z_1$ we derive a differential equation from (4.4) by comparing the coefficients of $\varepsilon$ and imposing the boundary condition $z_1 = 0$ for $t = t_0$. We simplify the calculation by introducing the functions $\omega$ and $w$ by

$$\left\{\begin{aligned} \omega &= \left(\zeta^2(t-t_0) + \frac{1}{4}\right)^{\frac{1}{4}} \\ w &= \omega z_1 \ . \end{aligned}\right. \tag{4.8}$$

Then the differential equation for $z_1$ takes the form

$$\frac{\dot{\zeta}}{\omega}\,\dot{w} = -\int_t^\infty \frac{\partial U_1}{\partial z}\,\dot{\zeta}dt\ ,$$

hence for $t = t_0$ we have

$$\dot{w}(t_0) = -\frac{1}{\sqrt{8}}\int_{t_0}^\infty \frac{\partial U_1}{\partial z}\,\dot{\zeta}dt \qquad \text{for} \qquad \varepsilon = 0\ .$$

Setting $t = t_0 + s$ and using (4.6) we get

$$\dot{w}(t_0) = +\frac{1}{\sqrt{8}}\,\frac{3}{4}\int_0^\infty \cos\,(t_0+s)\,\frac{\zeta\dot{\zeta}}{\omega^{10}}\,ds$$

where in $\zeta$, $\omega$ the argument is $s$. Using (4.7), (4.8) we find

$$\begin{aligned} \lambda(t_0,\varepsilon) &= 2 + \varepsilon\sqrt{2}\,\dot{w}(t_0) + O(\varepsilon^2) \\ &= 2 + \varepsilon\,\frac{3}{8}\,(-A\sin t_0 + B\cos t_0) + O(\varepsilon^2) \end{aligned} \tag{4.9}$$

where

$$A = \int_0^\infty \sin\,(s)\,\frac{\zeta\dot{\zeta}}{\omega^{10}}\,ds \tag{4.10}$$

and $B$ is a similar integral containing $\cos s$. Formula (4.9) shows that $\partial D_0$ is in first approximation an ellipse and our condition (4.2) for transversal intersection will be satisfied if $A > 0$.

c) To show $A > 0$ we use that $\zeta(s) > 0$, $\dot{\zeta}(s) > 0$ for $s > 0$, and since $\sin s > 0$ for $0 < s \leq \frac{\pi}{2}$ we have

$$A \geq \int_{\pi/2}^{\infty} (\sin s)(\zeta\dot{\zeta}\omega^{-10})ds = \int_{\pi/2}^{\infty} (\cos s)\frac{d}{ds}(\zeta\dot{\zeta}\omega^{-10})ds$$

where we applied integration by parts. Using the differential equation (4.5) for $\zeta$ we obtain the relation

$$\begin{aligned}\frac{d}{ds}(\zeta\dot{\zeta}\omega^{-10}) &= \ddot{\zeta}\zeta\omega^{-10} + \dot{\zeta}^2\frac{\partial}{\partial\zeta}(\zeta\omega^{-10}) \\ &= -\zeta\omega^{-6}\zeta\omega^{-10} + 2\omega^{-2}\frac{\partial}{\partial\zeta}(\zeta\omega^{-10}) \\ &= \omega^{-16}\{-9\zeta^2 + \tfrac{1}{2}\} = f(\zeta)\end{aligned}$$

which defines $f(\zeta)$. We will show that

$$\zeta^2(s) \geq \zeta^2\left(\frac{\pi}{2}\right) > 1 \quad \text{for} \quad s \geq \frac{\pi}{2} \tag{4.11}$$

so that $f(\zeta)$ is negative. By a standard argument we conclude from $f(\zeta) < 0$ that

$$\int_{\pi/2}^{\infty} \cos s\, f(\zeta)ds$$

is positive provided $f(\zeta)$ is a monotone decreasing function of $s$. Since $\dot{\zeta} > 0$ it suffices to show that

$$\frac{df}{d\zeta} = \frac{\partial}{\partial\zeta}\left(\omega^{-16}\left\{-9\zeta^2 + \frac{1}{2}\right\}\right) = \omega^{-20}\zeta\left(54\zeta^2 - \frac{17}{2}\right)$$

is positive for $s \geq \frac{\pi}{2}$, which is again a consequence of (4.11).

Finally, to verify (4.11) we use the monotonicity of $\zeta$. If (4.11) were false, i.e., if $\zeta\left(\frac{\pi}{2}\right) \leq 1$, then

$$\dot{\zeta}^2(s) = \frac{2}{\sqrt{\zeta^2 + \frac{1}{4}}} \geq \frac{4}{\sqrt{5}} > \frac{4}{3} \quad \text{for} \quad 0 \leq s \leq \frac{\pi}{2}$$

hence, with $\zeta(0) = 0$

$$\zeta^2\left(\frac{\pi}{2}\right) > \frac{4}{3} \cdot \left(\frac{\pi}{2}\right)^2 = \frac{\pi^2}{3} > 1$$

in contradiction to our assumption. This proves Lemma 3 completely.

5. *Proof of Lemma* 4 *of Chapter* III

a) We come to the study of the mapping $\phi$ near the boundary of $D_0$. This will be reduced to another mapping $\psi$ which we define now. We recall the differential equation (2.10) in the coordinates $x, y, t$ where the orbits parabolic for $t \to \pm\infty$ are given by $x = 0$ and $y = 0$, respectively. The solutions which do not escape lie in $x > 0$, $y > 0$ and with a small positive constant $a < 1$ we consider the flow in

(5.1) $$0 < x \leq a\ , \quad 0 < y \leq a\ , \quad 0 \leq t < 2\pi\ .$$

With another positive constant $\delta$ in

(5.2) $$0 < \delta \leq a < 1$$

we consider the annuli

(5.3) $$\begin{cases} A_0(\delta) : 0 < x < \delta\ , & y = a \\ A_1(\delta) : x = a\ , & 0 < y < \delta \end{cases}$$

in which $t$ is considered an angular variable. If $a$ is small enough the solutions of (2.10) passing through $A_j(a)$ intersect transversally and we will show that for sufficiently small $\delta$ the solutions starting at $A_0(\delta)$ pass through $A_1(a)$ as $t$ increases. By taking the first such intersection we define a mapping

(5.4) $$\psi : A_0(\delta) \to A_1(a)\ .$$

By following the orbits from $A_0(\delta)$ backwards to their first intersection with the plane $z = 0$ (notation of §2) for decreasing $t$ we define a mapping

$$\phi_- : A_0(\delta) \to D_0\ .$$

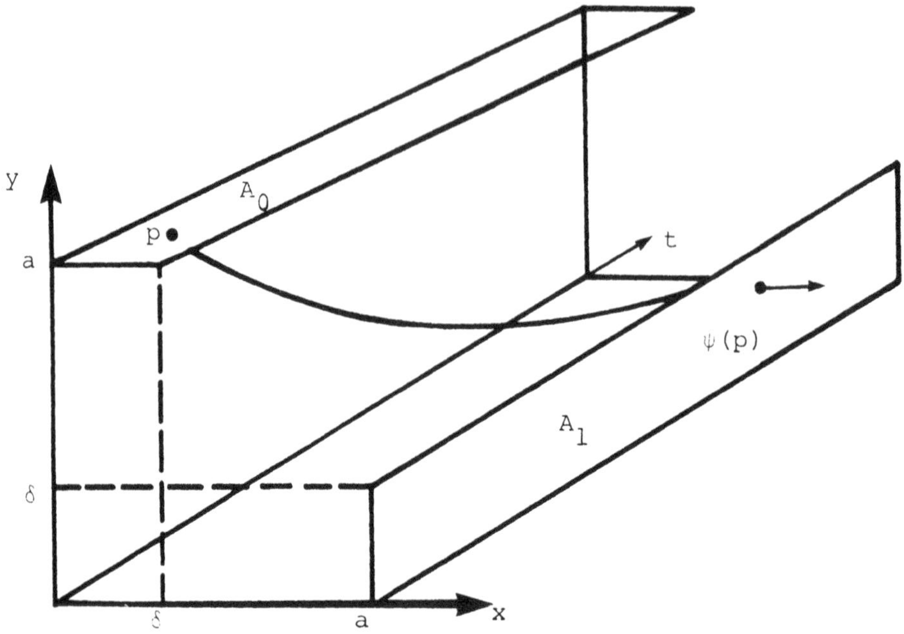


Fig. 20

Similarly, following the solutions from $A_1(a)$ forward to their first intersection with $z = 0$ we define

$$\phi_+ : A_1(a) \to D_1 \ .$$

These mappings are well defined and it is clear that our mapping $\phi$ is given by

$$\phi = \phi_+ \psi \phi_-^{-1} \tag{5.5}$$

if restricted to the boundary strip $\phi_-(A_0(\delta))$.

It is important to observe that $\phi_+$, $\phi_-$ are diffeomorphisms beyond the boundaries of $A_0$, $A_1$. For this reason the statements of Lemma 4 about $\phi$ will follow from similar statements about $\psi$. In particular, we will see that $\psi$ is not extendable continuously to the boundary $x = 0$ of $A_0$.

b) We denote the variables on $A_0(\delta)$ by $x_0$, $t_0$ and on $A_1(a)$ by $y_1$, $t_1$, so that $\psi$ takes $(x_0, t_0)$ into $(y_1, t_1)$. We show now: If $\alpha$ is a number in $0 < \alpha < \frac{1}{2}$. Then for sufficiently small $a$ and $\delta^{1-2\alpha} < a$ the mapping $\psi$ takes

$$\psi : A_0(\delta) \to A_1(\delta^{1-2\alpha}) \ . \tag{5.6}$$

This follows from some simple estimate for the solutions: We go back to the differential equation (2.8) considered in the domain (5.1). If $a$ is chosen small enough we have

$$(1-\alpha)x \le \frac{dx}{ds} \le (1+\alpha)x$$

and hence

$$x_0 e^{(1-\alpha)s} \le x(s) \le x_0 e^{(1+\alpha)s} \ . \tag{5.7}$$

Similarly, we find

$$y_0 e^{-(1+\alpha)s} \le y(s) \le y_0 e^{-(1-\alpha)s} \tag{5.8}$$

where we will choose $y_0 = a$. From (5.7) it follows that there exists an $s^*$ (depending on $(x_0, t_0)$) such that $x(s^*) = a$ where

$$\frac{1}{1+\alpha} \log \frac{a}{x_0} \le s^* \le \frac{1}{1-\alpha} \log \frac{a}{x_0} \ .$$

Thus, by (5.8)

$$y(s^*) \le a e^{-(1-\alpha)s^*} \le a\left(\frac{x_0}{a}\right)^{1-\alpha/1+\alpha} < \delta^{1-2\alpha} \tag{5.9}$$

since $\frac{1-\alpha}{1+\alpha} > 1-2\alpha$. This proves (5.4) for sufficiently small $\delta$. Furthermore we need an estimate for the time $T = t_1 - t_0$ required to go from $A_0(\delta)$ to $A_1(a)$. This $T$ is a function of $x_0$, $t_0$ and we claim: If $0 < \alpha < \frac{1}{2}$ and $a$ is small enough, then

$$T \ge x_0^{-\frac{3}{2}(1-2\alpha)} \tag{5.10}$$

hence $T \to \infty$ as $x_0 \to 0$.

To prove this we use the last equation of (2.8) and get, for sufficiently small a,

$$\frac{dt}{ds} \geq \frac{1}{4} (x+y)^{-3}$$

hence

$$T \geq \frac{1}{4} \int_0^{s^*} (x+y)^{-3} ds \ .$$

We define $s_0 > 0$ by

$$e^{2s_0} = \frac{y_0}{x_0} = \frac{a}{x_0}$$

so that $s_0 < s^*$ and

$$T \geq \frac{1}{4} \int_0^{s_0} (x+y)^{-3} ds \ .$$

In this integral we estimate x, y via (5.7), (5.8) and introduce the new independent variables $\sigma = s_0 - s$ and obtain after a short calculation

$$T \geq \frac{1}{4} \int_0^{s_0} (x_0 e^{(1+\alpha)s} + y_0 e^{-(1-\alpha)s})^{-3} ds$$

$$= \frac{1}{4} \left(\frac{x_0}{y_0}\right)^{\frac{3}{2}\alpha} (x_0\, y_0)^{-\frac{3}{2}} \int_0^{s_0} e^{+3\alpha\sigma}(e^{\sigma} + e^{-\sigma})^{-3} d\sigma \ .$$

Dropping $e^{3\alpha\sigma}$ in the integral and observing that the integral converges to a fixed number as $s_0 \to \infty$ we find

$$T \geq c_1 a^{-\frac{3}{2}(1+\alpha)} x_0^{-\frac{3}{2}(1-\alpha)} \geq c_1 x_0^{-\frac{3}{2}(1-\alpha)} \geq x_0^{-\frac{3}{2}(1-2\alpha)}$$

if $x_0$ is small enough, which gives (5.10).

c) To prove Lemma 4 we consider a $C^1$-arc in $A_0(\delta)$ abutting on $x_0 = 0$. We may represent it by $t_0 = t_0(x_0)$, $0 < x_0 < \delta$, where we may assume $|t_0(x_0)| \leq 2\pi$. Then we have for the image curve

$$t_1 = t_0 + T(x_0) \to \infty \quad \text{as} \quad x_0 \to 0 \ .$$

On the other hand by (5.9) $y_1 \to 0$ as $x_0 \to 0$ which shows that the image curve under $\psi$ spirals towards the boundary $y_1 = 0$ of $A_1$. Since in (5.5) the maps $\phi_+$, $\phi_-$ are differentiable including the boundary of the domain, the same property holds for $\phi$ proving Lemma 4.

## 6. *Proof of Lemma* 5 *of Chapter* III

### a) *Linearized equations*

The Lemma 5 deals with the Jacobian $d\phi$ of the map $\phi$ and we will accordingly study $d\psi$ for the map $\psi$ defined in the previous section. For this purpose we will consider the linearized equations associated with the differential equation (2.10)

$$\left\{\begin{aligned} \dot{\xi} &= f_x\xi + f_y\eta + f_t\tau \\ \dot{\eta} &= g_x\xi + g_y\eta + g_t\tau \\ \dot{\tau} &= 0 \end{aligned}\right. \tag{6.1}$$

and estimate its solutions for $t_0 \le t \le t_1$. One solution is trivially given by

$$\xi = f(x,y,t)\ , \quad \eta = g(x,y,t)\ , \quad \tau = 1 \tag{6.2}$$

where $x = x(t)$, $y = y(t)$ are solutions of (2.10). Our aim is to estimate two other solutions linearly independent of each other and of (6.2), for which $\tau = 0$. We observe that $\tau(t_0) = 0$ implies $\tau(t) = 0$ for $t > t_0$. By symmetry considerations it suffices to describe only one such solution which is done in the following lemma. We want to point out that the following technique to estimate solutions is not only useful for this particular example but can also be used with advantage in the study of flows or mappings near homoclinic solutions.

LEMMA 2. *Let* $\theta$ *be a constant in*

$$\tfrac{1}{2} < \theta < \sqrt{\tfrac{7}{9}} \tag{6.3}$$

*and the constant* a *in* (5.1) *small enough. Then one has for a solution of* (6.1) *the relations*

$$(6.4)\qquad |\eta(t)| \le \theta \sqrt{\frac{y(t)}{x(t)}}\, |\xi(t)| \ , \qquad \tau(t) = 0$$

*for* $t_0 \le t \le t_1$, *provided they hold initially, for* $t = t_0$. *Here* $x(t)$, $y(t)$ *are again solutions of* (2.10) *in* (5.1). *For any solution of* (6.1) *of this type one has, moreover,*

$$(6.5)\qquad |\xi(t_1)| \ge \sqrt{\frac{x(t_1)}{x(t_0)}}\, |\xi(t_0)| \ge \left(\frac{a}{\delta}\right)^{\frac{1}{2}} |\xi(t_0)| \ ,$$

*if* $0 < x(t_0) < \delta$.

*Proof*: The above estimates are derived very easily by introducing the variables u, v by

$$x = u^2 \ , \qquad y = v^2$$

where $0 < u < \sqrt{a}$, $0 < v < \sqrt{a}$. The differential equations (2.10) take the form

$$(6.6)\qquad \begin{cases} \dot u = F(u,v,t) = u((u^2 + v^2)^3 + O_8) \\ \dot v = G(u,v,t) = v(-(u^2 + v^2)^3 + O_8) \end{cases}$$

and the corresponding linearized differential equations

$$(6.7)\qquad \begin{aligned} \dot p &= F_u p + F_v q + F_t \tau \\ \dot q &= G_u p + G_v q + G_t \tau \\ \dot \tau &= 0 \ . \end{aligned}$$

Differentiating we see that

$$p = \frac{\xi}{2\sqrt{x}} \ , \qquad q = \frac{\eta}{2\sqrt{y}}$$

are solutions of (6.7) if $(x, y)$ are solutions of (2.10) and $\xi$, $\eta$ solutions of (6.1). Therefore our relations (6.4), (6.5) translate into

(6.8) $$|q(t)| \leq \theta|p(t)| \ , \quad \tau(t) = 0$$

provided they hold initially, and

(6.9) $$|p(t_1)| \geq |p(t_0)|$$

These relations are obviously much simpler and we shall verify them under the assumption (6.3).

We assume that (6.8) holds for $t = t_0$ and consider the expression $\theta^2 p^2(t) - q^2(t)$ which is positive for $t = t_0$. To show that this holds for all $t$ in $t_0 \leq t \leq t_1$ it suffices obviously to show that

(6.10) $$\frac{d}{dt}(\theta^2 p^2(t) - q^2(t)) \geq 0$$

when $\theta^2 p^2(t) - q^2(t)$ vanishes. Then this expression cannot change its sign. Using (6.7) and $\tau = 0$ we find

$$\frac{1}{2}\frac{d}{dt}(\theta^2 p^2 - q^2) = \theta^2(F_u p^2 + F_v pq) - G_u pq - G_v q^2$$

and if $\theta^2 p^2 = q^2$ this becomes

$$\frac{1}{2}\frac{d}{dt}(\theta^2 p^2 - q^2) = q^2(F_u \pm (\theta F_v - \theta^{-1} G_u) - G_v)$$

and with (6.6)

$$= q^2(u^2+v^2)^2\{7u^2 + v^2 \pm (\theta+\theta^{-1})6uv + u^2 + 7v^2 + O_4\}$$

$$= q^2(u^2+v^2)^2\{8(u^2+v^2) \pm (\theta+\theta^{-1})6uv + O_4\} \ .$$

The quadratic form in the parenthesis is indeed positive definite if

$$\theta + \theta^{-1} < \frac{8}{3}$$

and this is certainly true for $\frac{1}{2} < \theta < 1$. If $a$ is chosen small enough the error term $O_4$ is dominated and we have proven (6.10) and hence (6.8).

To prove (6.9) we deduce from (6.7) under the assumption (6.8) that

$$p\dot{p} = (F_u - \theta|F_v|)p^2$$

$$\geq (u^2+v^2)^2\{7u^2 + v^2 - \theta\, 6uv + O_4\}p^2$$

and the quadratic form on the right is positive definite if $\theta < \sqrt{\frac{7}{9}}$. Thus, under the assumption (6.3) we have, for a sufficiently small, that $p\dot{p} \geq 0$ which implies (6.9) by integration, and Lemma 2 is proven.

In the following the constant $\alpha$ in (5.6) will be fixed in

$$(6.11) \qquad 0 < \alpha < \frac{1}{3}$$

and a will be chosen so that the previous estimates are valid, while $\delta$ of (5.3) will be considered as a small parameter in $0 < \delta < a$, to be chosen later.

b) *Family of wedges*

To interpret the result of the above lemma we will restrict the initial values $x_0, y_0, t_0$ of the solution of (2.10) to the annulus $A_0(\delta)$ in (5.2) and choose $t_1$ as the first time $> t_0$ for which this solution meets $A_1(a)$ – or, according to (5.6) – $A_1(\delta^{1-2\alpha})$.* Then we consider the solutions of (6.1) along such a solution $x(t)$, $y(t)$ for $t_0 \leq t \leq t_1$. We will restrict the initial conditions $\xi_0, \eta_0, \tau_0$ to a wedge $W_0$ generated by

$$|\eta| \leq \theta\sqrt{\frac{a}{x_0}}\,|\xi|\ , \quad \tau = 0$$

and the vector given by the initial conditions

$$\xi = f_0 = f(x_0, a, t_0)\ , \quad \eta = g_0 = g(x_0, a, t_0)\ , \quad \tau = 1$$

* As in (5.9) we assume $\delta$ is chosen so small that $\delta^{1-2\alpha} < a$.

of the trivial solution. In other words, a vector $(\xi_0, \eta_0, \tau_0)$ belongs to $W_0$ if it can be written in the form

(6.12) $$\xi_0 = \xi^* + \lambda f_0 , \quad \eta_0 = \eta^* + \lambda g_0 , \quad \tau_0 = \lambda$$

with some real number $\lambda$ and with $\xi^*$, $\eta^*$ satisfying

(6.13) $$|\eta^*| \leq \theta \sqrt{\frac{a}{x_0}} \, |\xi^*| .$$

Observe, that (6.13) describes the sector (6.4) of the above lemma for $t = t_0$. According to the lemma the flow (6.1) takes such a wedge $W_0$ into another wedge $W_1$ generated by the sector

$$|\eta| \leq \theta \sqrt{\frac{y_1}{a}} \, |\xi|$$

and the vector $\xi = f_1 = f(a, y_1, t_1)$, $\eta = g_1 = g(a, y_1, t_1)$, $\tau = 1$. Thus, in order to obtain an assertion about the mapping $d\psi$ we merely have to intersect the wedges $W_0$, $W_1$ with the tangent planes of $A_0$, $A_1$, respectively. Denoting the tangent planes to $A_j$ by $T(A_j)(j = 0, 1)$ it follows that $d\psi$ takes the sector

$$W_0 \cap T(A_0) \quad \text{into} \quad W_1 \cap T(A_1)$$

and it remains to determine these sectors.

For a point $(\xi_0, \eta_0, \tau_0)$ in $W_0 \cap T(A_0)$ one has $\eta_0 = 0$, and therefore we have to take $\lambda = -\eta^* g_0^{-1}$ in (6.12), giving

(6.14) $$\xi_0 = \xi^* - \eta^* g_0^{-1} f_0 , \quad \eta_0 = 0 , \quad \tau_0 = -\eta^* g_0^{-1} ,$$

where $\xi^*$, $\eta^*$ satisfy (6.13). We diminish this sector slightly, and claim it contains the sector

(6.15) $$\mu_0 |\tau_0| \leq |\xi_0| , \quad \eta_0 = 0 \quad \text{where} \quad \mu_0 = f_0 + |g_0| \, \theta^{-1} \sqrt{\frac{x_0}{a}} .$$

Indeed (6.15) implies via (6.14)

$$\mu_0|\eta^*| = \mu_0|\tau_0 g_0| \le |\xi_0|\,|g_0| \le |\xi^*|\,|g_0| + |\eta^*|\,f_0$$

and since $\mu_0 > f_0$ it follows

$$|\eta^*| \le \frac{|g_0|}{\mu_0 - f_0}\,|\xi^*| = \theta\sqrt{\frac{a}{x_0}}\,|\xi^*|\ ,$$

which is (6.13). Thus (6.15) is contained in $W_0 \cap T(A_0)$.

Similarly, one verifies easily that $W_1 \cap T(A_1)$ is contained in the sector

(6.16) $$\xi_1 = 0\ ,\quad |\eta_1| \le \mu_1|\xi_1|$$

where

(6.17) $$\mu_1 = |f_1|\,\theta\sqrt{\frac{y_1}{x_1}} + |g_1|$$

and it follows that $d\psi$ takes the sector (6.15) into (6.16).

We introduce the notation

$$S_0'(\mu_0) : \mu_0|\tau_0| \le |\xi_0|$$

$$S_1(\mu_1) : |\eta_1| \le \mu_1\,|\tau_1|$$

so that $d\psi$ maps $S_0'(\mu_0)$ into $S_1(\mu_1)$ if $\mu_0$, $\mu_1$ are taken as in (6.15), (6.17). Observe that the sectors $S_0'(\mu_0)$ get smaller as $\mu_0$ increases while $S_1(\mu_1)$ get larger as $\mu_1$ increases and therefore the above assertion remains true if $\mu_0$, $\mu_1$ are both increased. With some positive constants $c_1$, $c_2$ depending on a, $\alpha$, but not on the solution or $\delta$ we find from (5.9), (2.10), (6.15) and (6.17)

$$\mu_0 \le c_1\sqrt{\delta}\ ,\quad \mu_1 \le c_2\delta^{\frac{1}{2}(1-2\alpha)}$$

and we conclude

(6.18) $$d\psi : S_0'(c_1\sqrt{\delta}) \to S_1\left(c_2\delta^{\frac{1}{2}(1-2\alpha)}\right).$$

Similarly, the estimate (6.5) leads to

$$|\tau_1| \geq c_3^{-1}\delta^{-\frac{1}{2}}|\xi_0| \tag{6.19}$$

for vectors $(\xi_0, 0, \tau_0) \in S_0'(c_1\sqrt{\delta})$, where $c_3$ is again a constant with the same qualifications as $c_1$, $c_2$ above. Indeed, if $(\xi_0, \eta_0, \tau_0) \in T(A_0)$ is written in the form (6.12), where now $\eta_0 = 0$ we have by (6.14)

$$|\xi^*| \geq |\xi_0| - |\tau_0| f_0 \geq \left(1 - c_1^{-1}\delta^{-\frac{1}{2}} f_0\right)|\xi_0| \geq \frac{1}{2}|\xi_0|$$

for sufficiently small $\delta$, since $f_0 = O(\delta)$.

If we write the image vector of $(\xi_0, \eta_0, \tau_0)$ with $\eta_0 = 0$ under $d\psi$ in the form $(\xi_1, \eta_1, \tau_1)$ with $\xi_1 = 0$, and set, analogously to (6.12)

$$0 = \xi_1 = \xi_1^* + \lambda_1 f_1 \ , \quad \eta_1 = \eta_1^* + \lambda_1 g_1 \ , \quad \tau_1 = \lambda_1 \tag{6.20}$$

we have by (6.5) in the Lemma 2

$$|\xi_1^*| \geq \left(\frac{a}{\delta}\right)^{\frac{1}{2}}|\xi^*| \geq \frac{1}{2}\left(\frac{a}{\delta}\right)^{\frac{1}{2}}|\xi_0|$$

and by (6.20)

$$|\tau_1| = f_1^{-1}|\xi_1^*| \geq (2a)^{-4}\,\frac{1}{2}\left(\frac{a}{\delta}\right)^{\frac{1}{2}}|\xi_0| = c_3^{-1}\delta^{-\frac{1}{2}}|\xi_0|$$

since $0 < f_1 < (2a)^4$. Thus (6.19) is verified with $c_3 = 2^5 a^{\frac{7}{2}}$.

In Figure 21 we indicated a "wide" sector $S_0'$ and a narrow sector $S_1$. It is clear that a curve $\gamma$, $t_0 = t_0(x_0)$, differentiable for $0 \leq x_0 \leq \delta$ is mapped into a curve spiralling towards the boundary $y_1 = 0$ on $A_1$. Moreover, for small $\delta$ the tangent vector of $\gamma$ will fall into $S_0'\left(c_1\delta^{\frac{1}{2}}\right)$ and hence by (6.18) the tangent of the image curve will lie in $S_1\left(c_2\delta^{\left(\frac{1}{2}\right)-\alpha}\right)$ showing that also the tangent of the image curve approaches that of $y_1 = 0$ as $x_0 \to 0$.

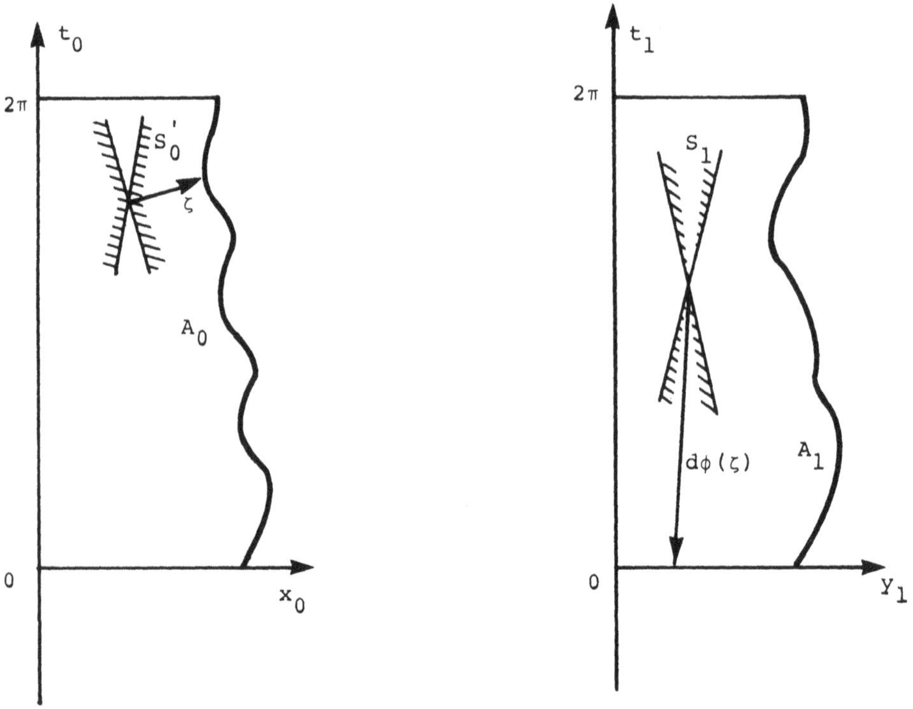


Fig. 21

From (6.19) we deduce a lower estimate for the derivative $T_{x_0}$ of the time $T(x_0, t_0) = t_1 - t_0$. For this purpose we consider the curve $\gamma : t_0 = \text{const.}$ which is mapped into $t_1 = t_0 + T(x_0, t_0)$, $y_1 = y_1(x_0, t_0)$. Since the tangent of $\gamma$ lies in $S_0\left(c_1 \delta^{\frac{1}{2}}\right)$ for $0 < x_0 \leq \delta$ we conclude from (6.19) that

$$\left|\frac{\partial T}{\partial x_0}\right| = \left|\frac{\partial t_1}{\partial x_0}\right| \geq c_3^{-1} \delta^{-\frac{1}{2}} \quad \text{for} \quad 0 < x_0 \leq \delta$$

hence

$$\left|\frac{\partial T}{\partial x_0}\right| \geq c_3^{-1} x_0^{-\frac{1}{2}} .$$

This estimate shows that also $|\frac{\partial T}{\partial x_0}|$ tends to infinity as $x_0 \to 0$. For T itself this was shown already in (5.10). It must be said, however, that the above estimate is very crude and does not even reflect the right exponent, but these estimates will suffice for our arguments.

c) *Completion of the proof of Lemma* 5

To finish the proof of Lemma 5 we merely have to use (5.5) to translate the estimates (6.18), (6.19) for $\psi$ into similar ones for $\phi$. Here we observe again that $\phi_+$, $\phi_-$ of (5.5) differentiable (in fact, real analytic) up to the boundary of $A_0(\delta)$, $A_1(\delta)$. Therefore, there are constants $c_4$, $c_5$ depending on the derivatives of $\phi_+$, $\phi_-$ such that

$$\phi_-(A_0(\delta)) \supset D_0(c_4^{-1}\delta)$$

and

$$\phi_+ : A_1(\delta) \to D_1(c_5\delta) \ .$$

Hence, setting $\delta_0 = c_4^{-1}\delta$ we conclude from (5.6) that, for sufficiently small $\delta_0$,

$$\phi : D_0(\delta_0) \to D_1(c_5(c_4\delta_0)^{1-2\alpha}) \subset D_1(\delta_0^{\beta})$$

if we take $\beta = 1-3\alpha$ which is positive by (6.11). This proves the first assertion of Lemma 5, if we replace $\delta_0$ by $\delta$.

Further it is clear, that $d\phi_-$ maps a tangent vector in the $t_0$-direction, at point $p \in A_0(\delta)$ into a vector at $\phi_-(p) \in D_0$ whose direction differs from that of the tangent vector of $\partial D_0$ at the point $q \in \partial D_0$, the closest boundary point to $\phi_-(p)$, by $O(\delta)$. For this one merely has to approximate $\phi_-$ by a map linear in $x_0$. Thus using the notation $\Sigma'_0$, $\Sigma_1$ of Chapter III, §5, d) and $S'_0$, $S_1$ of the previous section

$$d\phi_-\left(S'_0\left(c_1\delta^{\frac{1}{2}}\right)\right) \supset \Sigma'_0\left(c_5\delta_0^{\frac{1}{2}}\right) \supset \Sigma'_0\left(\delta_0^{\frac{1}{3}}\right)$$

with $\delta_0 = c_4^{-1}\delta$ sufficiently small. A similar consideration for $\phi_+$ gives

$$d\phi_+\left(S_1\left(c_2\delta^{\left(\frac{1}{2}\right)-\alpha}\right)\right) \subset \Sigma_1\left(c_6\delta_0^{\left(\frac{1}{2}\right)-\alpha}\right) \subset \Sigma_1\left(\delta_0^{\frac{\beta}{3}}\right)$$

since $\frac{1}{2} - \alpha > \frac{1}{2}(1-3\alpha) > \frac{1}{3}\beta$. Combining these estimates with (6.18) we get

$$d\phi : \Sigma_0'\left(\delta_0^{\frac{1}{3}}\right) \to \Sigma_1\left(\delta_0^{\frac{\beta}{3}}\right)$$

proving the second assertion of Lemma 5.

Finally, the last assertion follows in an obvious way from (6.19) and the observation that extra constants occurring from the different projections disappear since we lowered the exponent $\frac{1}{2}$ in (6.19) to $\frac{1}{3}$. Thus Lemma 5 is completely proven.

We remark that it clearly does not matter that we used orthogonal projections in Lemma 5 and one could equally well replace this by a projection with respect to directions transversal to the tangents of $\partial D_0$ or $\partial D_1$. This amounts to introduction of a different metric.

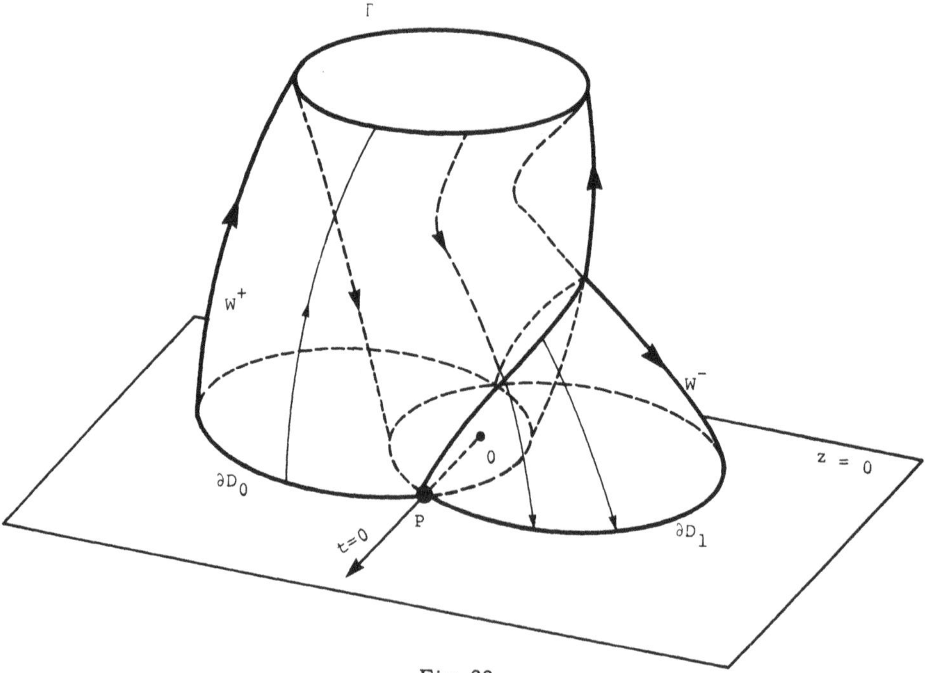


Fig. 22

In the figure we give a schematic view of the manifolds $W^+$, $W^-$ of orbits which are parabolic for $t \to +\infty$, and $t \to -\infty$, respectively. These are two-dimensional manifolds in the three-dimensional phase space. As the local study of §2 shows $W^+$ and $W^-$ intersect at infinity on a closed curve $\Gamma$, which we considered the analogue of a hyperbolic periodic orbit. But near $\Gamma$ these manifolds have no other points of intersection. On the other hand, $W^+$, $W^-$ meet the plane $z = 0$ in the curves $\partial D_0$, $\partial D_1$ respectively which intersect each other, in particular in $P$, a point playing the role of a homoclinic point.

## 7. *Proof of Theorem* 3.7, *concerning homoclinic points*

### a) *Local study of the hyperbolic point*

The proof of Theorem 3.7 will proceed quite analogously to that for the restricted three-body problem. We begin with the local study of the diffeomorphism $\phi$ near the hyperbolic fixed point $p$. We assume that the eigenvalues of the Jacobian at $p$ are $\lambda, \mu$ where

$$0 < \mu < 1 < \lambda . \tag{7.1}$$

It is well known that the invariant curves $W_p^+$, $W_p^-$ issuing from $p$ are $C^1$-curves; in fact, they are $C^\infty$-curves if $\phi$ is $C^\infty$ and analytic if $\phi$ is analytic. Staying in the class of $C''$-mappings we may introduce local coordinates, such that $p$ corresponds to $x = y = 0$, and the invariant curves to the coordinate axes. Denoting the image point of $x, y$ under $\phi$ by $(x_1, y_1)$ we have

$$\begin{aligned} x_1 &= f(x, y) \\ y_1 &= g(x, y) \end{aligned} \tag{7.2}$$

in a neighborhood of $x = y = 0$; here $f$, $g$ are continuously differentiable in this neighborhood and

$$\begin{aligned} f(0, y) &= g(x, 0) = 0 \\ f_x(0, 0) &= \lambda , \quad g_y(0, 0) = \mu . \end{aligned} \tag{7.3}$$

The first line of (7.3) expresses simply that $x = 0$ and $y = 0$ are invariant curves under $\phi$. We will restrict attention to the square

$$Q : 0 \leq x \leq a \ , \quad 0 \leq y \leq a \tag{7.4}$$

where $a > 0$ is chosen small enough. In the following we shall write $x_0$, $y_0$ for $x$, $y$, and denote its image point under $\phi^k$ by $x_k$, $y_k$.

With $\phi$ we associate the linearized mapping $d\phi$ given by

$$\xi_1 = f_x \xi + f_y \eta$$

$$\eta_1 = g_x \xi + g_y \eta \ .$$

The image point under $d\phi^k$ will be denoted by $\xi_k$, $\eta_k$.

LEMMA. *For sufficiently small* $a > 0$ *and any sequence of iterates* $(x_k, y_k)(k = 0, 1, \ldots, n)$ *in the interior of* $Q$ *the inequality*

$$|\eta_0| \leq \sqrt{\frac{y_0}{x_0}}\ |\xi_0|$$

*implies*

$$|\eta_k| \leq \sqrt{\frac{y_k}{x_k}}\ |\xi_k| \quad \textit{for} \quad k = 1, 2, \ldots, n \ . \tag{7.5}$$

*Moreover, under these assumptions one has*

$$|\xi_k| \geq \sqrt{\frac{x_k}{x_0}}\ |\xi_0| \ . \tag{7.6}$$

The proof uses the same trick as in §6: We introduce coordinates $u$, $v$ by $x = u^2$, $y = v^2$ so that our mapping has the form

$$u_1 = F(u, v) \ , \quad v_1 = G(u, v)$$

for $0 \leq u, v \leq \sqrt{a}$. Moreover, by (7.3)

$$F(0,v) \equiv G(u,0) = 0$$

$$F_u(0,0) = \sqrt{\lambda}\ , \quad G_v(0,0) = \sqrt{\mu}$$

and $F, G$ are continuously differentie e there, if $f, g \in C''$.* Writing the linearized mapping in the form

$$p_1 = F_u p + F_v q$$

$$q_1 = G_u p + G_v q$$

we find for $|q| \le |p|$

$$|p_1| \ge (\sqrt{\lambda} - O(a))|p|$$

$$|q_1| \le (\sqrt{\mu} + O(a))|p|$$

hence, if $a$ is sufficiently small

$$|q_1| \le |p_1|\ , \quad |p_1| \ge |p_0|\ .$$

More generally, we conclude

$$|q_k| \le |p_k| \quad \text{for} \quad k = 0, 1, \ldots, n$$

if this relation holds for $k = 0$ and provided that the points $(u_k, v_k)$ $(k = 0, \ldots, n)$ remain in $0 \le u, v \le \sqrt{a}$. Differentiation of $x = u^2$, $y = v^2$ shows that this is applicable to

$$p_k = \frac{\xi_k}{2\sqrt{x_k}}\ , \quad q_k = \frac{\eta_k}{2\sqrt{y_k}}$$

which yields the statement of the lemma.

---

* At this point we use that $\phi$ is a $C''$ diffeomorphism; this assumption can be replaced by milder ones. Working with the modulus of continuity of $f_x$, $f_y$, $g_x$, $g_y$ one can extend this argument to $C^1$-mappings.

This lemma expresses that in a sufficiently small square Q the sector bundle

$$|\eta| \le \sqrt{\frac{y}{x}}\ |\xi|$$

is mapped into itself under $d\phi$. Similarly one shows that

$$|\xi| \le \sqrt{\frac{x}{y}}\ |\eta|$$

is mapped into itself under $d\phi^{-1}$ provided the base points remain in Q.

Finally the inequality $|p_1| \ge |p|$ leads to $|p_{k+1}| \ge |p_k|$ which yields readily (7.6).

b) *Reduction to a local problem*

We will reduce the Theorem 3.7 to Theorem 3.2 and have to construct the set of vertical and horizontal strips satisfying the conditions i) and iii) of Chapter III, §3, §4. For this purpose we construct a quadrilateral R at the homoclinic point r, two of whose sides are parts of $W_p^+$, $W_p^-$ and the other two may be straight line segments, parallel to the tangents of $W_p^{\pm}$ at r. We reserve the freedom to make the sides of R small, and to choose R on the appropriate sides of $W_p^{\pm}$.

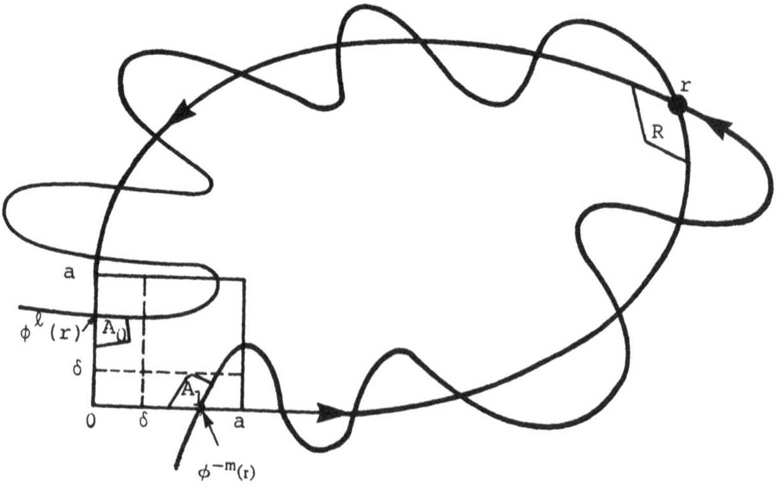


Fig. 23

Since r is a homoclinic point of $\phi$ so is $\phi^k(r)$ for every integer k. Moreover, there exists positive integers $\ell$, m such that

$$\phi^{\ell}(r) \in Q \ , \quad \phi^{-m}(r) \in Q \ .$$

We shall fix $\ell$, m as well as a for the following construction, which depends, however, on a small parameter $\delta$ to be chosen in

$$0 < \delta < a \ .$$

Furthermore, let b, c be defined in $0 < b, c < \frac{a}{2}$ such that $\phi^{\ell}(r)$ has the coordinates $(0, 2b)$ and $\phi^{-m}(r)$ the coordinates $(2c, 0)$.

Now we require that the sides of R are chosen so small that the image $\phi^{\ell}(R)$ lies not only in Q but in

$$0 \leq x \leq \delta \ , \quad b \leq y \leq a \tag{7.7}$$

and that $\phi^{-m}(R)$ in

$$c \leq x \leq a \ , \quad 0 \leq y \leq \delta \ . \tag{7.8}$$

If R is chosen on the appropriate sides of $W_p^{\pm}$ this is clearly possible, by continuity consideration, since $\ell$, m are fixed. The image domain

$$A_0 = \phi^{\ell}(R)$$

will then belong to (7.7) and one of its sides lies on $x = 0$ and one adjacent side will intersect $x = 0$ transversally. We diminish R further, if necessary, so that both sides adjacent to $x = 0$ can be written in the form $y = h_1(x)$ with a $C^1$-function $h_1$. Similarly, we may assume that the quadrilateral

$$A_1 = \phi^{-m}(R)$$

in (7.7) has two sides representable in the form $x = h_2(y)$ with a $C^1$-function $h_2$.

With this choice of $R$ and $A_0$, $A_1$ we define the transversal map $\psi$ from $A_0$, into $A_1$ naturally as follows: If $q \in A_0$ and there exists a $k > 0$ such that $\phi^k(q) \in A_1$ and $\phi(q), \phi^2(q), \ldots, \phi^{k-1}(q) \in Q$ then we say $q$ belongs to $D(\psi)$ and we set

$$\psi(q) = \phi^k(q) \in A_1$$

where $k$ is the smallest such $k > 0$. Thus $\psi$ is defined entirely in terms of the mapping (7.2) which was studied in a).

With these notations the mapping $\tilde{\phi}$ of Chapter III, §6 agrees with

$$\tilde{\phi} = \phi^m \psi \, \phi^\ell \tag{7.9}$$

in the common domain of definition.

c) *Construction of horizontal and vertical strips*

We claim that for the mapping $\psi$, defined in $D(\psi) \in A_0$, the range $\psi(D(\psi))$ intersects $A_1$ in infinitely many strips $U_k$, $(k = 1, 2, \ldots)$ whose horizontal boundaries are of the form

$$y = h(x)$$

where

$$\left|\frac{dh}{dx}\right| \leq \sqrt{\frac{\delta}{c}} \tag{7.10}$$

if $\delta$ is small enough; moreover, these strips connect the opposite sides of $A_1$.

For this we merely have to observe that, for sufficiently small $a$ we have in $Q$

$$x_1 \geq \lambda^{\frac{1}{2}} x \; , \quad y_1 \leq \mu^{\frac{1}{2}} y$$

so that the images under $\phi^k$ of the two curves $y = h_1(x)$, bounding $A_0$, will, for large $k$, intersect the domain (7.8) in a curve connecting $x = 0$ and $x = a$. Thus for large $k$ the domains $\phi^k(A_0) \cap A_1$ will consist of infinitely many quadrilaterals $\tilde{U}_k$ connecting opposite sides of $A_1$.

To obtain the additional information about the derivatives of the boundary curve we apply the above lemma to the tangent vectors of the boundary curves $y = h_1(x)$ of $A_0$. Since in $A_0$ we have $0 \leq x_0 \leq \delta$, $y_0 \geq b$ we can assure the condition

$$|\eta_0| \leq \sqrt{\frac{b}{\delta}}\ |\xi_0|$$

for those tangent vectors, which obviously is true if $\delta$ is made small enough. Thus, (7.5) implies for the image under $d\phi^k$

$$|\eta_k| \leq \sqrt{\frac{y_k}{x_k}}\ |\xi_k| \leq \sqrt{\frac{\delta}{c}}\ |\xi_k|$$

where we used that $(x_k, y_k)$ lies in (7.8). But this implies (7.10).

By the same consideration the pre-images $\psi^{-1}(\tilde{U}_k) = \tilde{V}_k$ are vertical strips in $A_0$ for which the derivatives $\frac{dx}{dy}$ of two boundary curves can be controlled, like in (7.10) (see Fig. 24). By definition we have

(7.11) $$\psi(\tilde{V}_k) = \tilde{U}_k$$

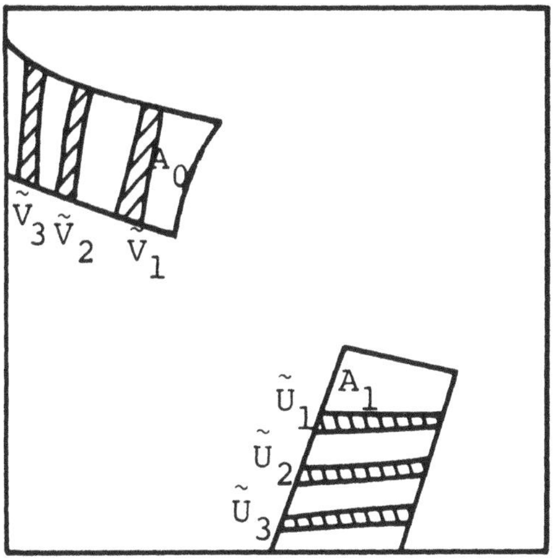


Fig. 24

Setting

$$U_k = \phi^m(\tilde{U}_k) , \quad V_k = \phi^{-\ell}(\tilde{V}_k)$$

we obtain a similar set of strips in R which by (7.11) and (7.9) satisfy

$$\tilde{\phi}(V_k) = \phi_m\psi(\tilde{V}_k) = U_k .$$

They play the role of the horizontal and vertical strips in Chapter III, §3 and §4 and the arguments of the previous section carry over to this one. In particular, the tangent direction of the boundaries of $U_k$, $V_k$ differs from those of one of the tangents of $W_p^{\pm}$ at r by at most $\delta^{\frac{1}{3}}$ as $\delta$ gets small. Thus the Lipschitz condition for these curves is easily verified. Moreover, the estimates (7.5), (7.6) lead to the conditions (4.3), (4.4), (4.5) of Chapter III, §4. These indications may suffice since the further details are straightforward.

## 8. *Nonexistence of integrals*

We supply the proof of Theorem 3.10. For this purpose we assume that $\phi$ is a diffeomorphism satisfying the hypotheses i) and iii) of Chapter III, §§3, 4. With these notations we saw that $\phi$ has an invariant set I, on which $\phi$ is equivalent to the shift $\sigma$. Under the additional hypothesis (4.14) of Chapter III, Theorem 3.4 implied that the points of I appeared as the points of intersection of a set of horizontal and vertical curves U(s), V(s) which are continuously differentiable.

We assume now that f is a continuous function on I satisfying

$$f(\phi(p)) = f(p) .$$

Since I contains a dense orbit, say $\phi^{(k)}(p_0)$ we have

$$f(\phi^k(p_0)) = f(p_0)$$

and, by continuity

$$f(q) = f(p_0) \quad \text{for all} \quad q \in I ,$$

i.e., $f$ is a constant on I. To construct such an orbit one merely has to form a sequence containing all blocks of sequences. This is possible since the set of such blocks is denumerable.

Now let $f \epsilon C^1(Q)$, and satisfy $f(\phi(p)) = f(p)$ on I. Then

$$\frac{\partial f}{\partial x} = \frac{\partial f}{\partial y} = 0 \quad \text{on} \quad I\ . \tag{8.1}$$

Indeed, if $p^* = \tau(s^*)$ is any point on I, then by Theorem 3.4, $p^*$ lies on the horizontal curve $U(s^*)$ and the vertical curve $V(s^*)$. By modifying the left tail of the sequence $s^*$ we may choose a sequence of points $q_\nu \epsilon U(s^*) \cap I$, $q \neq p^*$ converging to $p^*$. Thus if we parametrize $U(s^*)$ by $x$ we can express the directional derivative $D_1 f$ of $f$ along the tangent of $U(s^*)$ at $p^*$ as limit of a difference quotient of $f$ at $p^*$ and $q_\nu$. Since $f(p) = f(q_\nu)$ it follows

$$D_1 f = 0 \quad \text{at} \quad p^*\ .$$

By the same argument the directional derivative $D_2 f$ along the tangent of $V(s^*)$ vanishes. Since these directions are linearly independent we conclude (8.1).

If $f \epsilon C^\infty(Q)$ we can repeat this argument and deduce that all derivatives of $f$ vanish on I. Thus, if $f$ is real analytic it vanishes identically.

## BOOKS AND SURVEY ARTICLES

[1] ALEKSEEV, V. M., Sur l'allure finale du mouvement dans le problème des trois corps, Actes du Congres Int. des Math. 1970, vol. 2, 893-907, Gauthiers-Villars, Paris, 1971.

[2] ARNOLD, V. I., Small divisor problems in classical and celestial mechanics, Usp. Mat. Nauk *18*, no. 6 (114) 91-192 (1963).

[3] ARNOLD, V. I. and AVEZ, A., Problèmes ergodiques de la mécanique classique, Paris, 1967.

[4] BIRKHOFF, G. D., Dynamical Systems, AMS Coll. Publications, vol. 9, 1927, reprinted 1966.

[5] HAGIHARA, Y., Celestial Mechanics, M.I.T. Press, vols. I-V, 1970ff.

[6] KOLMOGOROV, A. N., Théorie générale des systèmes dynamiques et mécanique classique, Proc. Int. Congress of Math., Amsterdam, 1957. Translation in Appendix D of Abraham, R., Foundations of Mechanics, Benjamin, 1967.

[7] MOSER, J., Lectures on Hamiltonian Systems, Memoirs A.M.S. 81, 1968.

[8] SIEGEL, C. L. and MOSER, J. K., Lectures on Celestial Mechanics, Springer, Grundlehren Bd. 187, 1971.

[9] STERNBERG, S., Celestial Mechanics, vols. I, II, New York, 1969.

[10] SZEBEHELEY, V., Theory of Orbits, Academic Press, 1967.

[11] WINTNER, A., The Analytic Foundations of Celestial Mechanics, Princeton Univ. Press, 1947.

## CHAPTER I

[12] POINCARÉ, H., Sur la Stabilité du Système Solaire, Ouvres t. 8, pp. 538-547.

[13] KLEIN, F. and SOMMERFELD, A., Über die Theorie des Kreisels, Teubner, 1921, in particular, the discussion on stability and Laplace's stability proof of the planetary system on p. 342ff.

[14] MITTAG-LEFFLER, G., Zur Biographie von Weierstrass, Acta Math. *35*, 29-65, 1912, in particular, letter of Feb. 2, 1889, pp. 55-58 and letter to S. Kovalevski, 1878, p. 30.

[15] WEIERSTRASS, K. T. W., Über das Problem der Störungen in der Astronomie, Vorlesung, gehalten im mathematischen Seminar der Universität, Berlin, Wintersemester, 1880-81 (manuscript available at Mittag-Leffler Institute, Djursholm).

[16] BROUWER, D. and WOERKOM, A. J. J., The secular variations of the orbital elements of the principal elements, Astr. papers, Am. Ephemeris and Nautical Almanac *13*, pt. II, Washington, 1950, 81-107.

[17] GOLDREICH, P., An explanation of the frequent occurrence of commensurable mean motions in the solar system, Monthly Notices Roy. Astr. Soc. 130, 159-181 (1965).

[18] MOLCHANOV, A. M., The Resonant Structure of the Solar System, Icarus *8*, 203-215 (1968).

[19] BACKUS, G. E., Critique of "The Resonant Structure of the Solar System" by A. M. Molchanov, Icarus *11*, 88-92 (1969).

[20] HENON, M., A comment on "The Resonant Structure of the Solar System" by A. M. Molchanov, Icarus *11*, 93-94 (1969).

[21] MOLCHANOV, A. M., Resonances in Complex Systems: A reply to Critiques, Icarus *11*, 95-103 (1969).

[22] ARNOLD, V. I., On the classical perturbation theory and the stability problem of planetary systems, Dokl. Akad. Nauk SSSR *145*, 481-490 (1962).

[23] POINCARÉ, H., Sur le problème des trois corps et les équations de la dynamique, Acta Math. 13, 1-271 (1890).

[24] POINCARÉ, H., Les méthodes nouvelles de la mécanique céleste, vol. 2, Paris, 1893, esp. Sec. 148 and 149, pp. 99-105.

[25] ULAM, S. M., Proc. 4th Berkeley Symposium on Math. Statist. and Probability, Berkeley Univ. Cal. Press, *3*, p. 315 (1960).

[26] BRAHIC, A., Numerical study of a simple dynamical system, Astron. and Astrophys. *12*, 98-110 (1971).

[27] BRAUN, M., Particle Motion in a Magnetic Field, Journ. Diff. Eq. 8, 294-332 (1970).

[28] BRAUN, M., Structural stability and the Störmer problem, Ind. Univ. Math. J. 20, 469-497 (1970).

[29] BILLINGSLEY, P., Ergodic Theory and Information, Wiley and Sons, Inc., 1965.

[30] ANOSOV, D. V. and SINAI, Ja. G., Some smooth ergodic systems, Uspekhi Mat. Nauk 22, 107-172 (1967).

[31] FERMI, E., Beweis, dass ein mechanisches Normalsystem im allgemeinen quasi-ergodisch ist., Phys. Z. 24, 261-264 (1923).

[32] BIRKHOFF, G. D., On the periodic motions of dynamical systems, Acta Math. 50, 359-379 (1927), esp. footnotes on p. 379.

[33] SITNIKOV, K., Existence of oscillating motions for the three-body problem, Dokl. Akad. Nauk, USSR, *133*, no. 2, 303-306 (1960).

[34] ALEKSEEV, V. M., Quasirandom dynamical systems I, II, III, Math. USSR Sbornik *5*, 73-128 (1968); *6*, 505-560 (1968); *7*, 1-43 (1969).

[35] ALEKSEEV, V. M., Mat. Zametki *6* (4), 489-498, 1969.

[36] SMALE, S., Topology and Mechanics, Inv. Math. 10, 305-331 (1970) and 11, 45-64 (1970).

[37] EASTON, R., Some Topology of the 3-Body Problem, Journ. Diff. Eq. 10, 371-377 (1971).

[38] WEINSTEIN, A., Symplectic manifolds and their Lagrangean submanifolds, Adv. Math. 3, 329-349 (1971).

[39] WEINSTEIN, A., Lagrangean submanifolds and Hamiltonian systems, to appear.

[40] ARNOLD, V., Sur une propriété topologique des applications globalement canoniques de la mécanique classique, Compt. Rend. Acad. Sci., Paris, 261, 3719-3722 (1965).

[41] HARRIS, T. C., Periodic Solutions of Arbitrarily Long Periods in Hamiltonian Systems, Journ. Diff. Eq. 4, 131-141 (1968).

[42] ROBINSON, R. C., Generic properties of conservative systems I, II, Am. J. Math. *92*, 562-603, 897-906 (1970).

[43] SINAI, JA., Dynamical Systems with Countably Multiple Lebesgue Spectrum I, Am. Math. Soc. Transl. (2), 39, 83-110 (1964); II, Am. Math. Soc. Trans. (2), 68, 34-88 (1964).

[44] ORNSTEIN, D. S., Some new results in the Kolmogorov Sinai theory of entropy and ergodic theory, Bull. A.M.S. 77, 878-890 (1971).

## CHAPTER II

[45] LIAPOUNOFF, A., Problème général de la stabilite du mouvement, Ann. Fac. Sci. Toulouse (2), 203-474 (1907), reprinted, Ann. Math. Studies no. 17.

[46] BOCHNER, S. and MARTIN, W. T., Several Complex Variables, Princeton, 1948, esp. Chapter I; also references to C. Caratheodory, 1932, H. Cartan, 1932.

[47] SIEGEL, C. L., Über die analytische Normalform analytischer Differentialgleichungen in der Nähe einer Gleichgewichtslösung, Nachr. Akad. Wiss. Göttingen, Math. Phys. Kl., 21-30, 1952.

[48] SIEGEL, C. L., Iteration of analytic functions, Am. Math. 43, 607-612 (1942).

[49] POINCARÉ, H., Thèse, 1879, Oeuvres *1*, Paris, 1928.

[50] STERNBERG, S., The structure of local homeomorphisms II, III, Am. J. Math. 80, 623-632 (1958); 81, 578-604 (1959).

[51] HARTMAN, P., Ordinary Differential Equations, J. Wiley and Sons, Inc., 1964.

[52] PUGH, C. C., On a theorem of P. Hartman, Am. J. Math. 91, 363-367 (1969).

[53] BELLMAN, R., The iteration of power series in two variables, Duke Math. J. *19*, 339-347 (1952).

[54] NELSON, E., Topics in Dynamics I: Flows, Math. Notes, Princeton University Press, 1969.

[55] STERNBERG, S., Infinite dimensional Lie groups and formal aspects of dynamical systems, Journ. Math. Mech., 10, 451-474 (1961).

[56] SIEGEL, C. L., Über die Existenz einer Normalform analytischer Hamiltonischer Differentialgleichungen in der Nähe einer Gleichgewichtslösung, Math. Ann. 128, 144-170 (1954).

[57] BRJUNO, A. D., Analytical Forms of Differential Equations, Trudy Mosk. Mat. Obshest. 25, 119-262 (1971).

[58] GOLDSTEIN, H., Classical Mechanics, Addison-Wesley, 1950.

[59] ARNOLD, V. I., Proof of A. N. Kolmogorov's theorem on the preservation of quasi-periodic motions under small perturbations of the Hamiltonian, Usp. Mat. Nauk SSSR *18*, no. 5, 13-40 (1963).

[60] MOSER, J., On the construction of almost periodic solutions for ordinary differential equations, Proc. Int. Conf. on Functional Analysis and Related Topics, 60-67, Tokyo, 1969.

[61] BIRKHOFF, G. D., Proof of Poincaré's geometric theorem, Trans. A.M.S., 14, 14-22 (1913).

[62] BIRKHOFF, G. D., Surface transformations and their dynamical applications, Acta Math. 43, 1-119 (1920).

[63] MOSER, J., On invariant curves of area-preserving mappings of an annulus, Nachr. Akad. Wiss., Göttingen, Math. Phys. Kl., 1-20, 1962.

[64] OXTOBY, J., Measure and Category, Springer-Verlag, 1970.

[65] RÜSSMANN, H., Über invariante Kurven differenzierbarer Abbildungen eines Kreisringes, Nachr. Akad. Wiss., Göttingen II, Math. Phys. Kl., 67-105, 1970.

[66] MOSER, J., A rapidly convergent iteration method and nonlinear differential equations II, Ann. Acuola Normale Sup. Pisa, ser. III, *20*, 499-535 (1966).

[67] TAKENS, F., A $C^1$-counterexample to Moser's twist theorem, Indag. Math. 33, 379-386 (1971).

[68] DENJOY, A., Sur les courbes définies par les équations differentielles à la surface du tore, J. Math. Pures Appl. (9) 11, 333-375 (1933).

[69] ANOSOV, D. V. and KATOK, A. B., New examples in smooth ergodic theory, Ergodic Diffeomorphisms, Trudy Mosk. Math. Obsc., *23*, 3-36, 1970. Engl. transl. in Trans. Mosc. Math. Soc., Am. Math. Soc. 23, 1-35, 1972.

[70] BRJUNO, A. D., Instability in a Hamiltonian system and the distribution of asteroids, Mat. Sbornik *83* (125), no. 2, (1970); Math. USSR, Sbornik *12*, no. 2, pp. 272-312 (1970).

[71] JEFFERYS, W. H. and MOSER, J., Quasi-periodic solutions for the three-body problem, Astr. J. 71, 568-578 (1966).

[72] KRASINSKI, G. A., Quasi-periodic solutions of the first kind for the plane n-body problem, Trudy Inst. Theor. Astr. *13*, Leningrad, 105-168, 1969.

[73] KYNER, W. T., Rigorous and formal stability of orbits about an oblate planet, Mem. A.M.S., 81, 1968.

[74] LIEBERMAN, B., Existence of quasi-periodic solutions to the three-body problem, Celestial Mechanics 3, 408-426 (1971).

[75] LIEBERMAN, B., Quasi-periodic solutions of Hamiltonian systems, Journ. Diff. Eq., 11, 109-137 (1971).

[76] MOSER, J., Quasi-periodic solutions in the three-body problem, Bull Astr. (3) *3*, 53-59 (1968).

## CHAPTER III

[77] HADAMARD, J., Les surfaces à curbures opposés et leurs lignes géodesiques, Journ. de Math. (5) 4, 27-73 (1898).

[78] MORSE, M. and HEDLUND, G. A., Symbolic dynamics, Am. J. Math. 60, 815-866 (1938).

[79] LEVINSON, N., A second order differential equation with singular solutions, Ann. Math. 50, 127-153 (1949) (contains references to Cartwright-Littlewood and earlier work).

[80] SMALE, S., Diffeomorphisms with many periodic points, Differential and Combinatorial Topology (edited by S. S. Cairns) Princeton University Press, 63-80, 1965.

[81] SMALE, S., Differentiable Dynamical Systems, Bull. A.M.S. 73, 747-817 (1967).

[82] CONLEY, C. C., Low energy transit orbits in the restricted three-body problem, SIAM J. Appl. Math. *16*, 732-746 (1968).

[83] CONLEY, C. C., On the ultimate behavior of orbits with respect to an unstable critical point I, Oscillating, Asymptotic and Capture Orbits, J. Diff. Equ. 5, 136-158 (1969).

[84] CONLEY, C. and EASTON, R., Isolated invariant sets and isolating blocks, Trans. A.M.S. 158, 35-61 (1971).

[85] ARTIN, E., Ein mechanisches System mit quasiergodischen Bahnen, Abh. Math. Sem., Univ. Hamburg 3, 170-175 (1924).

[86] MOSER, J., On a theorem of Anosov, Journ, Diff. Eq. 5, 411-440 (1969).

[87] ALEKSEEV, V. M., On the capture orbits for the three-body problem for negative energy constant, Uspekhi Mat. Nauk 24, 185-186 (1969).

[88] HOPF, E., Ergodentheorie, Springer-Verlag, 1937, in particular, §13.

[89] BIRKHOFF, G. D., Nouvelles Recherches sur les systèmes dynamiques, Mem. Pont. Acad. Sci. Novi Lyncaei, 1, 85-216 (1935), in particular, Chapter IV.

[90] POINCARÉ, H., Les Méthodes Nouvelles de la Mécanique Céleste III, Gauthiers-Villars, 1899, esp. no. 395-404.

[91] ZEHNDER, E., Homoclinic points near elliptic fixed points, to appear in Comm. Pure Appl. Math. 1972.

## CHAPTER IV

[92] NISHIDA, T., A note on an existence of conditionally periodic oscillation in a one-dimensional anharmonic lattice, Mem. Fac. Eng., Kyoto Univ. 33, pt. 1, 27-34 (1971).

[93] LAX, P. D., Development of singularities of solutions of nonlinear hyperbolic partial differential equations, Journ. Math. Physics 5, 611-613 (1964).

[94] GARDNER, C., Kortweg-deVries equation and generalizations IV, Journ. Math. Phys. 12, 1548-1551 (1971).

[95] SAKHAROV, V. E. and FADDEEV, L. D., The Kortweg-deVries equation as an integrable Hamiltonian system, Funct. Anal. and its applications, *5*, 18-27 (1971).

## CHAPTER V

[96] NIRENBERG, L., An abstract form of the nonlinear Cauchy-Kowalewski theorem, to appear.

[97] SERGERAERT, F., Une généralisation du théorème des fonctions implicites de Nash, Compt. Rend. Acad. Sci., Paris, 270, 861-863 (1970).

[98] OSTROWSKI, A. M., Solutions of Equations and Systems of Equations, Academic Press, 1960.

[99] MOSER, J., Convergent series expansions for quasi-periodic motions, Math. Ann. 169, 136-176 (1967).

[100] GRAFF, S., On the conservation of hyperbolic invariant tori for Hamiltonian systems, Dissertation at New York University, June 1971.

## CHAPTER VI

[101] SLOTNICK, D. L., Asymptotic behavior of solutions of canonical systems near a closed, unstable orbit, Contribution to the Theory of Nonlinear Oscillations IV, Ann. Math. Studies 41, 85-110 (1958).

[102] McGEHEE, R., A stable manifold theorem for degenerate fixed points with applications to Celestial Mechanics, to appear in Journ. Diff. Eq.

Milton Keynes UK
Ingram Content Group UK Ltd.
UKHW030942140924
448309UK00002B/108

9 780691 089102